"十三五"高等职业教育规划教材

iOS开发基础教程
（Swift版）

iOS KAIFA JICHU JIAOCHENG（Swift BAN）

陈志峰　吴道君　主　编
翟高粤　田　英　张晓云　副主编

中国铁道出版社有限公司
CHINA RAILWAY PUBLISHING HOUSE CO., LTD.

内 容 简 介

本书主要用于 iOS 移动应用开发技术的学习,全书共分 6 章,内容包括:认识 iOS 开发、编写第一个应用程序、听声音识动物、找出"你我"不同、组建平面图形乐队以及创建机器人聊天室等。本书根据苹果公司较新的 Swift 3 语言,结合当前开发手机应用的必要工具,针对 iOS 开发所需要的声音、动画和图形处理等移动应用,最后通过学习网络技术,实现与机器人在线聊天功能。

本书从实际应用出发学习 Swift 语言,有别于一般教材以语法为主干组织内容的模式,更符合学习者获取知识的规律。

本书相对完整地介绍了 iOS 开发的新知识、新技术、新方法和新应用,较好地满足了学习者对 iOS 开发基础技术和基本技能训练的需要。

本书适合作为高等职业学校移动应用开发(iOS)课程的教材,也可作为 iOS 开发爱好者的自学参考书。

图书在版编目(CIP)数据

iOS 开发基础教程:Swift 版/陈志峰,吴道君主编. —北京:中国铁道出版社,2017.9(2019.8重印)
"十三五"高等职业教育规划教材
ISBN 978-7-113-23430-0

Ⅰ.①i… Ⅱ.①陈… ②吴… Ⅲ.①移动终端-应用程序-程序设计-高等职业教育-教材 Ⅳ.①TN929.53

中国版本图书馆 CIP 数据核字(2017)第 194289 号

书　　名	iOS 开发基础教程(Swift 版)
作　　者	陈志峰　吴道君　主编

策　　划	汪　敏	读者热线	(010)63550836
责任编辑	李丽娟　彭立辉		
封面设计	刘　颖		
责任校对	张玉华		
责任印制	郭向伟		

出版发行	中国铁道出版社有限公司(100054,北京市西城区右安门西街 8 号)
网　　址	http://www.tdpress.com/51eds/
印　　刷	三河市航远印刷有限公司
版　　次	2017 年 9 月第 1 版　2019 年 8 月第 3 次印刷
开　　本	787 mm×1 092 mm　1/16　印张:13.75　字数:274 千
书　　号	ISBN 978-7-113-23430-0
定　　价	42.00 元

版权所有　侵权必究

凡购买铁道版图书,如有印制质量问题,请与本社教材图书营销部联系调换。电　话:(010)63550836
打击盗版举报电话:(010)51873659

前言

本书主要用于 iOS 移动应用开发技术的学习。苹果 iOS 是由苹果公司开发的移动设备操作系统，2016 年 3 月，库克在春季发布会上宣布目前全球活跃的 iOS 设备达到 10 亿台，iOS 已经成为世界最大的移动操作系统之一。开发者可以将自己的应用上传至 App Store 中，供用户下载使用。目前，每年 App Store 下载量都在持续增长，然而面对如此广阔的市场，国内的 iOS 开发人才却非常稀缺。

本书所采用的开发语言是 Swift 3。相较于传统的 Objective-C 语言来说，Swift 的代码更简洁，开发效率更高，涵盖了现在流行的编程方式，如结构化、面向对象、泛型、函数式等。随着时间推移，Swift 在整个苹果移动应用开发中所占的代码比例会越来越大。

本书注重教学规律和课堂教学，相对完整地介绍了使用 Swift 语言进行 iOS 开发的新知识、新技术、新方法和新应用，较好地满足了学习者对 iOS 开发基础技术和基本技能训练的需要。

教材特色

➢ 循序渐进：按照学生学习、认知过程，由浅入深地组织教学内容。

➢ 讲解到位：对关键技术的理论知识讲解细致，采用直观的图解方式，便于教师教学和学生自学。

➢ 注重实践：在通过实例学习基本理论和编程知识后，结合音频图像处理和网络技术等，对所学技术进行深层了解，以提高读者的职业竞争力和创新力。

教材内容

全书共分 6 章，内容包括：认识 iOS 开发、编第一个应用程序、听声音识动物、找出"你我"不同、组建平面图形乐队以及创建机器人聊天室等。本书根据苹果公司较新的 Swift 3 语言，结合当前开发手机应用的必要工具，针对 iOS 开发所需要的声音、动画和图形处理等移动应用，通过学习网络技术，实现与机器人在线聊天功能。

教学安排

序号	章节名称	知识点	建议讲授课时	建议实验课时
1	认识 iOS 开发	苹果操作系统及系列产品	1	
2		苹果开发语言介绍	1	
3		Swift 语言简介	1	1
4	编写第一个应用程序	Swift 开发环境搭建	1	2
5		编写第一个应用程序	1	1
6		Outlet 和 Action	1	2
7		Playground 的使用	1	4
8	听声音识动物	开发规划	1	
9		音频播放	1	1
10		动画播放	1	1
11		功能实现	2	4
12		拓展学习	1	2
13	找出"你我"不同	开发规划	1	
14		功能实现	1	4
15	组建平面图形乐队	开发规划	1	
16		初始图形世界	1	1
17		触摸事件和绘图	1	1
18		绘图类介绍及应用	1	1
19		功能实现	1	4
20	创建机器人聊天室	开发规划	1	
21		表视图	1	2
22		图灵机器人 API	1	2
23		网络访问方式	1	1
24		功能实现	2	6
	合　计		26	40

配套资源

　　本书提供了内容丰富的配套资源,包括电子课件、源程序、素材以及案例库、题库、素材库等多项辅助内容,读者可以登录出版社网址 http://www.tdpress.com/51eds/,输入书本名称或 ISBN 号查找本书,进行配套资源下载。

适合读者

- 希望从事 iOS 编程的开发人员。
- 具有一定语言基础,希望进一步提高技能的人员。
- 高等职业院校相关专业的学生和老师。
- 即将毕业的大学毕业生。
- 相关培训机构的学员和老师。
- 对 iOS APP 开发有兴趣的编程爱好者。

本书作者

本书由陈志峰、吴道君任主编,翟高粤、田英、张晓云任副主编,卢爱红、陆萍、岳健和王庆承等参与编写。其中:第 2、6 章由陈志峰编写,第 4 章由吴道君编写,第 5 章由翟高粤编写,第 3 章由田英编写,第 1 章由张晓云编写,卢爱红、陆萍、岳健和王庆承参与了例题程序的调试和习题的编写。张波、常玉祥、胡莹婷、庄朋杰、赵莹莹、张海玉等同学也参与了教学项目的调试,在此表示感谢。

由于高职计算机教育发展迅速,加之编者水平有限,疏漏和不足之处在所难免,我们会不断总结经验,及时修订和完善本教材。感谢读者使用本书,不足之处敬请广大读者朋友批评指正,欢迎大家提出宝贵意见。

编 者
2017 年 5 月

目 录

第 1 章　认识 iOS 开发 ··· 1
1.1　macOS 和 iOS ·· 1
1.2　苹果系列产品 ·· 3
1.3　Objective-C ·· 7
1.4　Cocoa Touch ·· 10
1.5　Swift 语言简介 ·· 14

第 2 章　编写第一个应用程序 ································· 29
2.1　Swift 开发与学习环境 ································ 29
2.2　第一个 iOS 应用程序 ································ 32
2.3　Outlet 和 Action ·· 54
2.4　Playground 环境 ·· 67

第 3 章　听声音识动物 ··· 73
3.1　功能简介 ·· 73
3.2　音频播放 ·· 75
3.3　动画播放 ·· 82
3.4　功能实现 ·· 84
3.5　拓展学习:纯代码编程 ································ 89

第4章 找出"你我"不同 99
4.1 功能简介 99
4.2 故事板(Storyboard) 100

第5章 组建平面图形乐队 115
5.1 初识图形世界 116
5.2 触摸事件和绘图 124
5.3 拥有自己的绘图类 129
5.4 绘制平面几何图形 139
5.5 奏响乐队凯歌 147

第6章 创建机器人聊天室 165
6.1 表视图(UITableView) 165
6.2 图灵机器人 API 177
6.3 网络访问 URLSession 183
6.4 基于表格的聊天界面 191

附录 A 用户界面要素 199

附录 B iOS 俱乐部护照 207

第 1 章

认识 iOS 开发

iOS 是 2007 年由苹果公司专门为 iPhone 手机开发的移动操作系统,要开发 iOS 应用程序,就需要 Mac 计算机、macOS 操作系统和 Xcode 开发环境,并熟悉 Objective-C 或者 Swift 语言。

1.1 macOS 和 iOS

macOS 是苹果公司(www.apple.com)于 2016 年 6 月为 Macintosh 系列计算机开发的操作系统,最新版本为 macOS Sierra,如图 1-1 所示。

图 1-1 苹果 macOS Sierra 中版本信息对话框

2001年，苹果公司推出了当时业界全新的PC操作系统Mac OS X，并从2002年起随Macintosh计算机发售，获得了巨大的成功。它是一套基于UNIX的操作系统，包含两个主要的部分：核心名为Darwin，是以FreeBSD源代码和Mach微核心为基础，由苹果公司和独立开发者社区协力开发；一个由苹果公司开发，名为Aqua的专有版权的图形用户界面。

Mac OS X Server亦同时于2001年发售，架构上来说与工作站（客户端）版本相同，只有在包含的工作组管理和管理软件工具上有所差异，［如邮件传输服务器、Samba软件、LDAP目录服务器，以及名称服务器（DNS）］，提供对于关键网络服务的简化访问。同时它也有不同的授权型态。

Mac OS X包含了一系列软件开发工具，其重大的特色是名为Xcode的集成开发环境。Xcode是一个能与数种编译器沟通的接口，包括C、C++、Objective-C和Java，最新支持Swift语言，可以编译出目前Mac OS X所运行的硬件平台上的可执行文件。

macOS是由Mac OS X升级而来的，其系统菜单如图1-2所示。

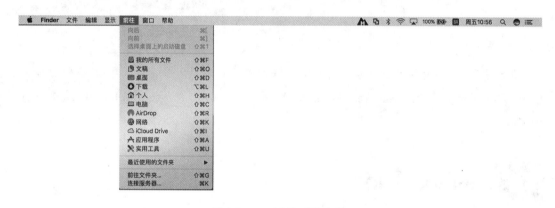

图1-2　macOS的系统菜单

macOS Sierra主要功能包括：

（1）Auto Unlock：可以使用iPhone和Apple Watch解锁Mac，当靠近计算机时，无须输入密码，打开计算机即可自动进入桌面。

（2）Universal Clipboard：自动同步在不同设备上的剪贴板记录，可以很流畅地在手机上复制计算机上的内容。

（3）视频播放支持画中画模式，悬浮于所有窗口之上，可自由拖动和调节窗口大小，与iPad上的效果一致。网页上的HTML5视频原生支持该模式，其他视频播放器需要添加新的API代码以支持该模式。

（4）Launchpad可让用户迅速访问应用程序，只需点击Dock内的Launchpad图标即可。可以通过访问Mac App Store，登录相应的Apple ID下载应用程序，下载完成后将自动出现在Launchpad上，随时待用，如图1-3所示。

iOS是由苹果公司开发的移动操作系统。苹果公司最早于2007年1月9日的Macworld大会上公布这个系统，最初是设计给iPhone使用的，后来陆续套用到iPod touch、iPad以及

Apple Watch 等产品上。iOS 与 Mac OS X 操作系统一样,属于类 UNIX 的商业操作系统。原本这个系统名为 iPhone OS,因为 iPad、iPhone、iPod touch 都使用 iPhone OS,所以 2010 WWDC 大会上宣布改名为 iOS,其界面如图 1-4 所示。

图 1-3　macOS 的 Dock

图 1-4　iPhone 上的 iOS 界面

1.2　苹果系列产品

1. iPhone

iPhone 是美国苹果公司研发的智能手机,搭载了 iOS 操作系统。2007 年 1 月 9 日,乔布斯在旧金山马士孔尼会展中心的苹果公司全球软件开发者年会首次推出第一代 iPhone。两个最初型号分别是售价为 499 美元的 4 GB 和 599 美元的 8 GB 版本,于当地时间 2007 年 6 月 29 日下午 6 时在美国正式发售。由于刚推出的 iPhone 上市后引发热潮,销情反应热烈,被部分媒体誉为"上帝手机"。

2017 年 3 月 21 日,苹果推出了一款拥有动人红色外观的特别版 iPhone 7 和 iPhone 7 Plus。图 1-5 所示为部分 iPhone 系列手机。

图 1-5　部分 iPhone 系列手机

2. iPad

iPad 是由苹果公司于 2010 年开始发布的平板计算机系列，由英国出生的设计主管 Jonathan Ive 领导的团队设计。iPad 定位介于苹果的智能手机 iPhone 和笔记本式计算机产品之间，与 iPhone 布局一样（屏幕中有 4 个虚拟程序固定栏），提供浏览网站、收发电子邮件、观看电子书、播放音频或视频、玩游戏等功能，如图 1-6 所示。

图 1-6　iPad 系列平板计算机

苹果公司在 iPad 上最新推出了 Swift Playground，为儿童和青少年在用 Swift 语言编程提供便利。图 1-7 所示为 Swift Playground 编程环境。

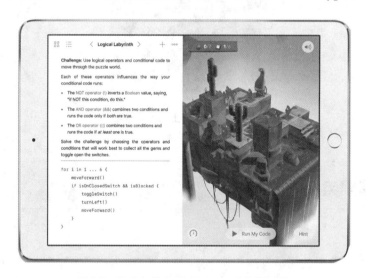

图 1-7　iPad 上的 Swift Playground 编程环境

3. Apple Watch

Apple Watch 是由苹果公司 CEO Tim Cook 于 2014 年 9 月 9 日在加州库比蒂诺的 Flint 表演艺术中心公布的一款智能手表。2015 年 3 月 10 日，苹果在旧金山 Moscone Center 召开 2015 年春季新品发布会，正式发布了 Apple Watch，分为 Apple Watch、Apple Watch Sport 和 Apple Watch Edition 三个系列。

用户使用最新的 Watch OS 操作系统,可以即时运行应用程序,更换多种炫酷的界面,和朋友共享活动等。Siri 可以通过语音识别来帮助用户做更多的工作,Breathe 等应用程序有助于用户的健康,如图 1-8 所示为 Apple Watch 系列智能手表。

图 1-8　Apple Watch 系列智能手表

4. iMac

iMac 是一款针对消费者和教育市场的一体化苹果 Macintosh 计算机系列。iMac 的特点是它的设计。早在 1998 年,苹果总裁史蒂夫·乔布斯就将 What's not a computer"(不是计算机的计算机)概念应用于设计 iMac 的过程。结果造就了软糖——iMac G3、台灯——iMac G4 和像框——iMac G5。由于 iMac 在设计上的独特之处和出众的易用性,它几乎连年获奖。如图 1-9 所示为一款 iMac 计算机。

每台新购买的 Mac 均配备照片、iMovie、GarageBand、Pages、Numbers 和 Keynote。让用户从开启它的那一刻起,就能尽情挥洒创意。同时,用户还可享有多款精彩 APP,用于收发电子邮件、畅游网络、发送文本信息、进行

图 1-9　iMac 计算机

FaceTime 视频通话,甚至还有一款专门的 APP,能够帮助用户寻找更多新的 APP。图 1-10 所示为苹果计算机所提供的各种 APP。

5. Mac mini

Mac mini 由苹果公司设计,是 Mac 产品线的一员,于 2005 年 1 月 11 日的 Macworld 上公布。其低价、小巧、易用的设计,吸引了很多用户。Mac mini 最初推出了两款不同的型号,两者于 2005 年 1 月 22 日在美国推出(1 月 29 日全球发售);2006—2012 年,Mac Mini 屡次更新;而 2012 年之后推出的 Mac mini,几乎都采用了 Intel Core i5 及 Intel Core i7 处理器。和其他桌面式计算机相比,占用空间更小,能源消耗更少。Mac mini 拥有精致的铝合金机身以及清爽的银色外表,小巧、雅致、落落大方。它看起如此简单,以至于用户很难把它和计算机联系在一起。Mac mini 不包括键盘、鼠标和显示器,其外观和背板如图 1-11 所示。

图 1-10 苹果计算机所提供的各种 APP

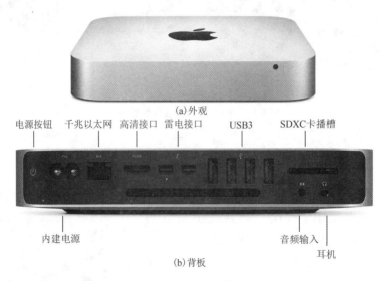

图 1-11 Mac mini 计算机外观及其背板

6. MacBook

MacBook 是一款配备强劲的处理器、先进的图形功能和高速内存的苹果笔记本式计算机。2015 年 3 月 9 日,苹果春季发布会在美国旧金山芳草地艺术中心召开,重点发布了 MacBook 12 英寸新机型。MacBook 采用了全新的 USB-C 接口,更为简洁。USB-C 全称为 USB Type-C,属于 USB 3.0 下一代接口,其亮点在于更加纤薄的设计、更快的传输速率(最高可达 10 Gbit/s)、更强的电力传输(最高 100 W),此外,USB-C 接口还支持双面插入,正反面随便插,相比 USB 2.0/USB 3.0 更为先进。

图 1-12 MacBook 笔记本计算机

MacBook 采用全新设计,分为灰、银、金三色,12 英寸 Retina 显分辨率为 2 304×1 440 像素,处理器为英特尔酷睿 M 低功耗处理器,如图 1-12 所示。它采用无风扇设计,这也是首台无风扇的 MacBook。新 MacBook 质量约 0.91 kg,厚 13.1 mm,比 11 英寸的 MacBook Air 薄 24%。主板比之前版本小了 67%。触控板的压力传感器能检测到用户在面板上用了多大的力,使用创新的阶梯式电池,使电池容量大大提高。

MacBook Pro 是苹果公司用来取代 PowerBook G4 产品线的英特尔核心的笔记本式计算机。2006 年 1 月 11 日由该公司首席执行官史蒂夫·乔布斯在 MacWorld 2006 大会上发布,并已于该年 2 月正式出货。MacBook Pro 与新的 iMac(酷睿)同为第一款转换为英特尔核心的产品。Intel Based 苹果计算机延续了之前经典的"咚"启动音。

2016 年 10 月 28 日,加入 Touch Bar 的新款 MacBook Pro 发布,如图 1-13 所示。

MacBook Air 是苹果公司在 2008 年 2 月 19 日推出的当时世界上最薄的笔记本式计算机。MacBook Air 具有 11.6 英寸和 13.3 英寸两个尺寸,被美国知名媒体《商业内幕》在 2013 年度纳入"本年度最具创新力的十大设备"。

MacBook Air 之所以能做到如此之薄,主要源自于 LED 屏幕和特殊处理器的采用,MBA 所采用的处理器是英特尔专门为苹果定制的,这种定制的处理器也属于酷睿 2 系列,但是面积比标准的酷睿 2 处理器要小很多,功耗也要低不少。

2010 年 10 月,苹果发布第二代 MacBook Air,如图 1-14 所示。这次升级使 MacBook Air 有了两种机型:传统的 13.3 英寸和新增的 11.6 英寸。它们的最大特点是用闪存代替硬盘,64~256 GB 的闪存被直接嵌入主板,节省了巨大空间来存放电池,这使 13.3 英寸机型的使用时间提升到 7 小时。

2016 年 4 月,苹果公司更新了 MBA,内存提高 8 GB,售价基本未变。2017 年,全线 MBA 产品更新到 8 GB 内存。

用户开发 iOS 一般建议购买 MacBook 或者 MacBook Pro 笔记本式计算机。

图 1-13 MacBook Pro 笔记本式计算机

图 1-14 MacBook Air 笔记本式计算机

1.3 Objective-C

Objective-C,通常写作 ObjC 或 OC,是在 C 语言基础上扩充的面向对象的编程语言,用于开发 macOS 和 iOS 应用程序。如图 1-15 所示为 Objective-C 中的 ViewController.m 文件。

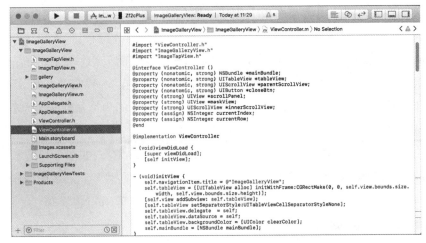

图 1-15 Objective-C 中的 ViewController.m 文件

20 世纪 80 年代初,布莱德·考克斯(Brad Cox)在其公司 Stepstone 发明 Objective-C。他对软件设计和编程里的真实可用度问题十分关心。

GCC 与 Clang 含 Objective-C 的编译器,Objective-C 可以在 GCC 以及 Clang 运作的系统上编译。

Objective-C 是 C 语言的严格超集,任何 C 语言程序不经修改就可以直接通过 Objective-C 编译器,在 Objective-C 中使用 C 语言代码也是完全合法的。Objective-C 被描述为盖在 C 语言上的薄薄一层,因为 Objective-C 的原意就是在 C 语言主体上加入面向对象的特性。表 1-1 所示为 Objective-C 代码的文件扩展名。

表 1-1 Objective-C 代码的文件扩展名

扩 展 名	内 容 类 型
.h	头文件,包含类、类型、函数和常数的声明
.m	典型的源代码文件扩展名,可以包含 Objective-C 和 C 代码
.mm	源代码文件。带有这种扩展名的源代码文件,除了可以包含 Objective-C 和 C 代码以外,还可以包含 C++ 代码。仅在 Objective-C 代码中确实需要使用 C++ 类或者特性的时候才用这种扩展名

Objective-C 的面向对象语法源于 Smalltalk 消息传递风格。所有其他非面向对象的语法,包括变量类型、预处理器(preprocessing)、流程控制、函数声明与调用皆与 C 语言完全一致。图 1-16 所示为一个简单的 Objective-C 源程序。

```
#import <Foundation/Foundation.h>
int main(int argc, char * argv[]) {
    @autoreleasepool {
        NSLog(@ "Hello World!");
    }
    return 0;
}
```

图 1-16 一个简单的 Objective-C 源程序

Objective-C 最大的特色是承自 Smalltalk 的消息传递模型(Message Passing),此机制与现在的 C++ 的主流风格有很大差异。在 Objective-C 中,与其说对象互相调用方法,不如说对象之间互相传递消息更为精确。此 2 种风格的主要差异在于调用方法/消息传递这个动作。C++ 中类与方法的关系严格清楚,一个方法必定属于一个类,在编译时(Compile Time)就已经绑定,不可能调用一个不存在类中的方法。但在 Objective-C 中,调用方法视为向对象发送消息,所有方法都被视为对消息的回应。所有消息处理直到运行时(Runtime)才会动态决定,并交由类自行决定如何处理收到的消息。也就是说,一个类不保证一定会响应收到的消息,如果类收到了一个无法处理的消息,程序只会抛出异常,不会出错或崩溃。

类是 Objective-C 用来封装数据及操作数据的行为的基础结构。对象就是类的运行期间实例,它包含了类声明的实例变量自己的内存复制以及类成员的指针。Objective-C 的类规格说明包含了两部分:接口(Interface)与实现(Implementation)。接口部分包含了类声明和实例变量的定义以及类相关的方法;实现部分包含了类方法的实际代码。

C++ 调用一个方法的语法为:obj.method(argument);而 Objective-C 则写成[obj method:argument]。

以一个 MyClass 的类定义为例,这个类继承自 NSObject 基础类。类声明总是由@interface 编译选项开始,由@end 编译选项结束,如图 1-17 所示。类名之后的(用冒号分隔的)是父类的名字。类的实例(或者成员)变量声明在被大括号包含的代码块中。实例变量块后面就是类声明的方法的列表。每个实例变量和方法声明都以分号结尾。类的定义文件遵循 C 语言之惯例以 .h 为扩展名,实现文件以 .m 为扩展名。方法前面的 +/- 号代表函数的类型:加号(+)代表类方法(Class Method),不需要实例就可以调用,与 C++ 的静态函数(Static Member Function)相似;减号(-)即是一般的实例方法(Instance Method)。

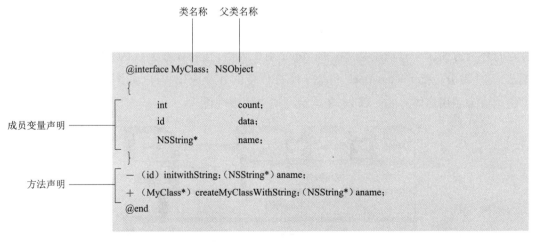

图 1-17 Objective-C 中类的声明

Objective-C 创建对象需通过 alloc 以及 init 两个消息。alloc 的作用是分配内存,init 则是初始化对象。init 与 alloc 都是定义在 NSObject 中的方法,父对象收到这两个信息并做出正确回应后,新对象才创建完毕。

```
MyObject* my=[[MyObject alloc] init];
```

Objective-C 中的类可以声明两种类型的方法:实例方法和类方法。实例方法就是一个方法,它在类的一个具体实例的范围内执行。也就是说,在调用一个实例方法前,必须首先创建类的一个实例。而类方法不需要创建一个实例。

方法声明包括方法类型标识符、返回值类型、一个或多个方法标识关键字、参数类型和名信息。以图 1-18 的 insertObject:atIndex:方法的声明为例,这个方法有两个参数。方法声明由一个减号(-)开始,这表明这是一个实例方法。方法实际的名字(insertObject:atIndex:)是所有方法标识关键字的级联,包含了冒号。冒号表明了参数的出现。如果方法没有参数,可以省略第一个(也是唯一的)方法标识关键字后面的冒号。

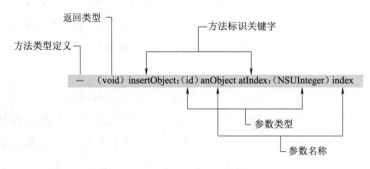

图 1-18　方法的声明

1.4　Cocoa Touch

Cocoa API 是 Mac OS X 的应用程序标准,主要包括两方面:运行环境方面和开发方面。在运行环境方面,Cocoa 应用程序呈现 Aqua 用户界面,且和操作系统的其他可视部分紧密集成,这些部分包括 Finder、Dock 和基于所有环境的其他应用程序,如图 1-19 所示。Cocoa 无缝地成为了用户体验的一部分,在运行环境方面表现优秀。

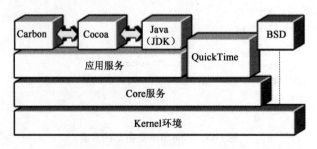

图 1-19　Cocoa 在 Mac OS X 中的位置

但是,程序员更感兴趣的是开发方面。Cocoa 是一个面向对象的软件组件/类的集成套件,它使开发者可以快速创建健壮和全功能的 Mac OS X 应用程序。这些类是可复用和可支配的软件积木,开发者可以直接使用,或者根据具体需求对其进行扩展。从用户界面对象到网络,几乎每个想象得到的开发需求都存在对应的 Cocoa 类;对于没有预想到的需求,也可以轻松地从现有类派生出子类来实现。

在各种面向对象的开发环境中,Cocoa 有着最为著名的血统。从 1989 年作为 NeXTSTEP 推出到现在,人们一直对它进行精化和测试。它优雅而强大的设计完美地适合所有类型的快速软件开发:不仅适合开发应用程序,也适合开发命令行工具、插件和不同类型的程序包。Cocoa 为应用程序提供很多行为和外观,使程序员有更多的时间用于特色功能的开发上。

核心的 Cocoa 类库封装在两个框架中:Foundation 和 Application Kit 框架。同所有框架一样,这两个框架不仅包含动态共享库(有时是几个兼容版本的库),还包含头文件、API 文档和相关的资源。Application Kit 和 Foundation 框架的分割反映了 Cocoa 编程接口分为图形用户界面部分和非图形接口。

iOS 开发框架相当于目录,在这个目录包含了共享库、通过访问共享库的头文件和其他图片声音等资源,被应用程序所调用。这些框架构成了 iOS 操作系统的层次架构,一共分为 4 层,从上到下依次为:Cocoa Touch Layer(触摸 UI 层)、MediaLayer(媒体层)、Core Services Layer(核心服务层)、Core OS Layer(核心 OS 层)。

Cocoa Touch 是从 Cocoa 发展而来,专门用于 iOS 开发,常见的 UIKit 和 Foundation 就位于其中,如图 1-20 所示。

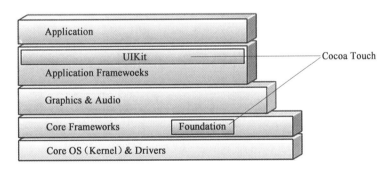

图 1-20　UIKit 和 Foundation 在 Cocoa Touch 中的位置

Foundation 框架(Foundation.framework)定义了大量的类,大体可分为:数值和字符串;数组、字典和集合、操作系统服务;处理日期和时间、内存管理、文件系统、URL、进程通信、通知、归档和序列化、处理几何数据结构(如点和长方形),如图 1-21 所示。

UIKit 框架(UIKit.framework)包含 iOS 中实现图形、事件驱动编程等关键架构 Objective-C 的编程接口,如图 1-22 所示。在图中可以看出,Responder 类是最大分支的根类,UIResponder 为处理响应事件和响应链定义了界面和默认行为。当用户用手指滚动到表或者在虚拟键盘上输入时,UIKit 就生成时间传送给 UIResponder 响应链,直到链中有对象处理这个事件。相应的核心对象，如 UIApplication、UIWindow 和 UIView 都直接或间接地从 UIResponder 继承。

UIKit 实际是 Mac OS 中的 Cocoa 的 AppKit 的变种。

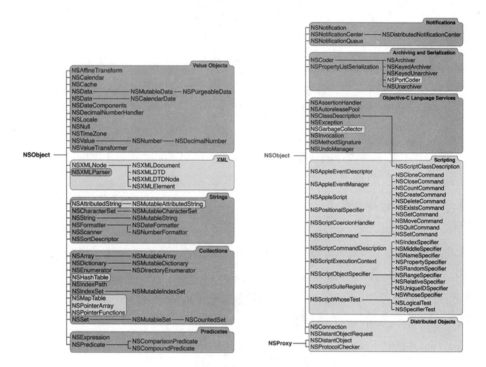

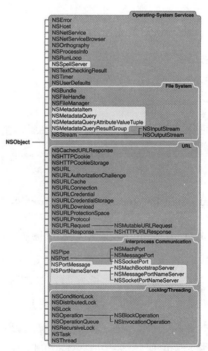

图 1-21　Foundation 库的主要类和 API

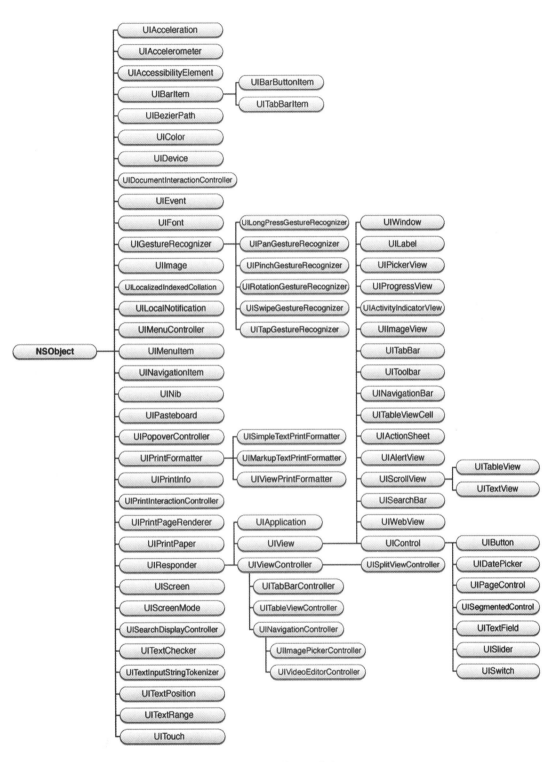

图 1-22 UIKit 库的主要类和 API

1.5 Swift 语言简介

Swift 是苹果公司于 2014 年 6 月 3 日在 WWDC 2014 上发布的一门全新编程语言。Swift 语言是用来编写 MacOS 和 iOS 应用程序的语言，建立在 C 语言和 Objective-C 语言基础之上，没有 C 语言的兼容性限制，采用安全模型的编程架构模式，使整个编程过程更加容易和灵活，并且完全支持主流框架：Cocoa 和 Cocoa Touch 框架，现在 Swift 语言已经开源（www.swift.org）。

Swift 的开发结合了众多工程师的心血，同时也借鉴了 Objective-C、Rust、Ruby 等其他语言的优点。其核心吸引力在于 Swift 语言的交互式版本 Xcode Playgrounds 和实时代码预览 REPL 功能。Read-Eval-Print-Loop（"读取-求值-输出"-"循环"，简称 REPL）：通过实时运行新编写的代码，立即获得其结果，是一种类似脚本的运行环境。

Swift 使用安全的编程模式并添加了很多新特性，这将使编程更简单、扩展性更强，也更有趣。Swift 支持 Cocoa 和 Cocoa Touch 框架，通过改进编译器、调试器和框架结构，让 Swift 使用自动引用计数（Automatic Reference Counting，ARC）来简化内存管理。

Swift 对于初学者来说也很简单。Swift 是一门既满足工业标准又像脚本语言一样充满表现力和趣味的编程语言。Swift 支持代码预览，这个革命性的特性可以允许程序员在不编译和运行应用程序的前提下运行 Swift 代码并实时查看结果。

Swift 提供了 C 和 Objective-C 的所有基础数据类型，包含整型 Int，浮点型 Double、Float，布尔型 Bool 以及 String 字符串。同时，Swift 也提供了两种强大的集合数据类型：Array（数组）和 Dictionary（字典）。

另外，Swift 引入了在 Objective-C 中没有的一些高级数据类型，例如 Tuples（元组），可以创建和传递一组数值。

Swift 还引入了可选项类型（Optionals），用于处理变量值不存在的情况。可选项的意思有两种：一是变量是存在的，例如等于 X；二是变量值根本不存在。Optionals 类似于 Objective-C 中指向 nil 的指针，但是适用于所有的数据类型，而非仅仅局限于类。Optionals 相比于 Objective-C 中 nil 指针更加安全和简明，并且也是 Swift 诸多最强大功能的核心。

Swift 语言以一只灵动的雨燕为代表符号，如图 1-23 所示。

图 1-23 Swift 语言的标志

1.5.1 简单值

Swift 语言使用 let 来声明常量，使用 var 来声明变量。一个常量的值，在编译的时候，并不需要有明确的值,但是只能为它赋值一次。也就是说,可以用常量来表示这样一个值:只需要决定一次,但是需要使用很多次。例如:

```
print("Hello,world")   //向控制台输出字符串"Hello,world"
var myVariable = 42
myVariable = 50
let myConstant = 42
```

常量或者变量的类型必须和赋给它们的值一样。但是,不用明确地声明类型,如果声明的同时赋值,编译器会自动推断类型。在上面的例子中,编译器推断出 myVariable 是一个整数(Integer),因为它的初始值是整数。

如果初始值没有提供足够的信息(或者没有初始值),就需要在变量后面声明类型,用冒号分割。例如:

```
let implicitInteger = 70
let implicitDouble = 70.0
let explicitDouble:Double = 70
```

练习:创建一个常量,显式指定类型为 Float 并指定初始值为 4。

值永远不会被隐式地转换为其他类型。如果需要把一个值转换成其他类型,需显式转换。例如:

```
let label = "The width is"
let width = 94
let widthLabel = label + String(width)
```

练习:删除最后一行中的 String,错误提示是什么?

有一种更简单的把值转换成字符串的方法:把值写到括号中,并且在括号之前写一个反斜杠。例如:

```
let apples = 3
let oranges = 5
let appleSummary = "I have \(apples) apples. "
let fruitSummary = "I have \(apples + oranges) pieces of fruit. "
```

练习:使用\()来把一个浮点计算转换成字符串,并加上某人的名字,和他打个招呼。

使用方括号[ ]来创建数组和字典,并使用下标或者键(key)来访问元素。最后一个元素后面允许有个逗号。

```
var shoppingList = ["catfish", "water", "tulips", "blue paint"]
shoppingList[1] = "bottle of water"
var occupations = [
    "Malcolm":"Captain",
    "Kaylee":"Mechanic",
]
occupations["Jayne"] = "Public Relations"
```

要创建一个空数组或者字典，使用初始化语法。

```
let emptyArray = [String]()
let emptyDictionary = [String:Float]()
```

如果类型信息可以被推断出来，可以用 [ ] 和 [:] 来创建空数组和空字典——就像声明变量或者给函数传参数的时候一样。

```
shoppingList = []
occupations = [:]
```

1.5.2 控制流

使用 if 和 switch 来进行条件操作，使用 for...in、for、while 和 repeat...while 来进行循环。包裹条件和循环变量括号可以省略，但是语句体的大括号是必需的。

```
let individualScores = [75, 43, 103, 87, 12]
var teamScore = 0
for score in individualScores {
    if score > 50 {
        teamScore += 3
    } else {
        teamScore += 1
    }
}
print(teamScore)
```

在 if 语句中，条件必须是一个布尔表达式——这意味着像 if score {...} 这样的代码将报错，而不会隐形地与 0 做对比。

通常可以一起使用 if 和 let 来处理值缺失的情况。这些值可由可选值来代表。一个可选的值是一个具体的值或者 nil 以表示值缺失。在类型后面加一个问号来标记这个变量的值是可选的。

```
var optionalString:String? = "Hello"
print(optionalString == nil)

var optionalName:String? = "John Appleseed"
var greeting = "Hello!"
if let name = optionalName {
    greeting = "Hello, \(name)"
}
```

练习：把 optionalName 改成 nil，greeting 会是什么？添加一个 else 语句，当 optionalName 是 nil 时给 greeting 赋一个不同的值。

如果变量的可选值是 nil，条件会判断为 false，大括号中的代码会被跳过。如果不是 nil，会将值赋给 let 后面的常量，这样代码块中就可以使用这个值。

另一种处理可选值的方法是通过使用"??"操作符来提供一个默认值。如果可选值缺失的,可以使用默认值来代替。

```
let nickName:String? = nil
let fullName:String = "John Appleseed"
let informalGreeting = "Hi \(nickName ?? fullName)"
```

switch 支持任意类型的数据以及各种比较操作——不仅仅是整数及测试相等。例如:

```
let vegetable = "red pepper"
switch vegetable {
case "celery":
    print("Add some raisins and make ants on a log.")
case "cucumber", "watercress":
    print("That would make a good tea sandwich.")
case let x where x.hasSuffix("pepper"):
    print("Is it a spicy \(x)?")
default:
    print("Everything tastes good in soup.")
}
```

练习:删除 default 语句,看看会有什么错误。

注意 let 在上述例子的等式中是如何使用的,它将匹配等式的值赋给常量 x。

运行 switch 中匹配到的子句之后,程序会退出 switch 语句,并不会继续向下运行,所以不需要在每个子句结尾写 break。

可以使用 for...in 来遍历字典,需要两个变量来表示每个键值对。字典是一个无序的集合,所以它们的键和值以任意顺序迭代结束。例如:

```
let interestingNumbers = [
    "Prime":[2, 3, 5, 7, 11, 13],
    "Fibonacci":[1, 1, 2, 3, 5, 8],
    "Square":[1, 4, 9, 16, 25],
]
var largest = 0
for(kind, numbers)in interestingNumbers {
    for number in numbers {
        if number > largest {
            largest = number
        }
    }
}
print(largest)
```

练习:添加另一个变量来记录最大数字的种类,同时仍然记录这个最大数字的值。

使用 while 来重复运行一段代码直到不满足条件。循环条件也可以在结尾,保证能至少循环一次。例如:

```
var n = 2
while n < 100 {
    n = n * 2
```

```
}
print(n)

var m = 2
repeat {
    m = m * 2
} while m < 100
print(m)
```

可以在循环中使用"..<"来表示范围。例如：

```
var total = 0
for i in 0..<4 {
    total += i
}
print(total)
```

使用"..<"创建的范围不包含上界，如果想包含的话需要使用"..."。

1.5.3 函数

使用 func 来声明一个函数，使用名字和参数来调用函数。使用" -> "来指定函数返回值的类型。例如：

```
func greet(name:String, day:String) -> String {
    return "Hello \(name), today is \(day)."
}
greet("Bob", day:"Tuesday")
```

练习：删除 day 参数，添加一个参数来表示今天吃了什么午饭。

使用元组来让一个函数返回多个值，该元组的元素可以用名称或数字来表示。例如：

```
func calculateStatistics(scores:[Int]) ->(min:Int, max:Int, sum:Int){
    var min = scores[0]
    var max = scores[0]
    var sum = 0
    for score in scores {
        if score > max {
            max = score
        } else if score < min {
            min = score
        }
        sum += score
    }
    return(min, max, sum)
}
let statistics = calculateStatistics([5, 3, 100, 3, 9])
print(statistics.sum)
print(statistics.2)
```

函数可以带有可变个数的参数，这些参数在函数内表现为数组的形式。例如：

```
func sumOf(numbers:Int...) -> Int {
    var sum = 0
    for number in numbers {
        sum += number
    }
    return sum
}
sumOf()
sumOf(42, 597, 12)
```

练习：写一个计算参数平均值的函数。

函数可以嵌套，被嵌套的函数可以访问外侧函数的变量，一般不可以使用嵌套函数来重构一个太长或者太复杂的函数。例如：

```
func returnFifteen() -> Int {
    var y = 10
    func add(){
        y += 5
    }
    add()
    return y
}
returnFifteen()
```

函数是可以嵌套，这意味着函数可以作为另一个函数的返回值。例如：

```
func makeIncrementer() ->(Int -> Int){
    func addOne(number:Int) -> Int {
        return 1 + number
    }
    return addOne
}
var increment = makeIncrementer()
increment(7)
```

函数也可以当作参数传入另一个函数。例如：

```
func hasAnyMatches(list:[Int], condition:Int -> Bool) -> Bool {
    for item in list {
        if condition(item){
            return true
        }
    }
    return false
}
func lessThanTen(number:Int) -> Bool {
    return number < 10
}
var numbers = [20, 19, 7, 12]
hasAnyMatches(numbers, condition:lessThanTen)
```

1.5.4 对象和类

使用 class 和类名来创建一个类。类中属性的声明和常量、变量声明一样，唯一的区别就是它们的上下文是类。同样，方法和函数声明也一样。例如：

```
class Shape {
    var numberOfSides = 0
    func simpleDescription() -> String {
        return "A shape with \(numberOfSides) sides."
    }
}
```

练习：使用 let 添加一个常量属性，再添加一个接收一个参数的方法。

要创建一个类的实例，可在类名后面加上括号。使用"."来访问实例的属性和方法。例如：

```
var shape = Shape()
shape.numberOfSides = 7
var shapeDescription = shape.simpleDescription()
```

这里的 Shape 类缺少了一些重要的东西：一个构造函数来初始化类实例。使用 init 来创建一个构造器。例如：

```
class NamedShape {
    var numberOfSides: Int = 0
    var name: String
    init(name: String) {
        self.name = name
    }
    func simpleDescription() -> String {
        return "A shape with \(numberOfSides) sides."
    }
}
```

注意：self 被用来区别实例变量。当创建实例时，像传入函数参数一样给类传入构造器的参数。每个属性都需要赋值——无论是通过声明（就像 numberOfSides）还是通过构造器（就像 name）。

如果需要在删除对象之前进行一些清理工作，可使用 deinit 创建一个析构函数。

子类的定义方法是在它们的类名后面加上父类的名字，用冒号分割。创建类时并不需要一个标准的根类，所以可以忽略父类。

子类如果要重写父类的方法，需要用 override 标记——如果没有添加 override 就重写父类方法，编译器会报错。编译器同样会检测 override 标记的方法是否确实在父类中。例如：

```
class Square: NamedShape {
    var sideLength: Double
    init(sideLength: Double, name: String) {
        self.sideLength = sideLength
```

```
            super.init(name:name)
            numberOfSides = 4
        }
        func area() ->  Double {
            return sideLength* sideLength
        }
        override func simpleDescription() -> String {
            return "A square with sides of length \(sideLength). "
        }
}
let test = Square(sideLength:5.2, name:"my test square")
test.area()
test.simpleDescription()
```

创建 NamedShape 的另一个子类 Circle,构造器接收两个参数:一个是半径,另一个是名称。在子类 Circle 中实现 area( )和 simpleDescription( )方法。例如:

```
class EquilateralTriangle:NamedShape {
    var sideLength:Double = 0.0
    init(sideLength:Double, name:String){
        self.sideLength = sideLength
        super.init(name:name)
        numberOfSides = 3
    }
    var perimeter:Double {
        get {
            return 3.0* sideLength
        }
        set {
            sideLength = newValue / 3.0
        }
    }
    override func simpleDescription() -> String {
        return "An equilateral triagle with sides of length \(sideLength). "
    }
}
var triangle = EquilateralTriangle(sideLength:3.1, name:"a triangle")
print(triangle.perimeter)
triangle.perimeter = 9.9
print(triangle.sideLength)
```

除了存储简单的属性之外,属性还可以有 getter 和 setter 。

在 perimeter 的 setter 中,新值的名字是 newValue。可以在 set 之后显式地设置一个名字。

注意:EquilateralTriangle 类的构造器执行了三步:

(1)设置子类声明的属性值。

(2)调用父类的构造器。

(3)改变父类定义的属性值。

其他的工作比如调用方法 getters 和 setters 也可以在这个阶段完成。

如果不需要计算属性,但是仍然需要在设置一个新值之前或者之后运行代码,使用 willSet 和 didSet。

例如,下面的类确保三角形的边长总是和正方形的边长相同。

```
class TriangleAndSquare {
    var triangle:EquilateralTriangle {
        willSet {
            square.sideLength = newValue.sideLength
        }
    }
    var square:Square {
        willSet {
            triangle.sideLength = newValue.sideLength
        }
    }
    init(size:Double, name:String){
        square = Square(sideLength:size, name:name)
        triangle = EquilateralTriangle(sideLength:size, name:name)
    }
}
var triangleAndSquare = TriangleAndSquare(size:10, name:"another test shape")
print(triangleAndSquare.square.sideLength)
print(triangleAndSquare.triangle.sideLength)
triangleAndSquare.square = Square(sideLength:50, name:"larger square")
print(triangleAndSquare.triangle.sideLength)
```

处理变量的可选值时,可以在操作(比如方法、属性和子脚本)之前加"?"。如果"?"之前的值是 nil,"?"后面的参数都会被忽略,并且整个表达式返回 nil。否则,"?"之后的参数都会被运行。在这两种情况下,整个表达式的值也是一个可选值。

```
let optionalSquare:Square? = Square(sideLength:2.5, name:"optional square")
let sideLength = optionalSquare?.sideLength
```

1.5.5 枚举和结构体

使用 enum 创建一个枚举。就像类和其他所有命名类型一样,枚举可以包含方法。例如:

```
enum Rank:Int {
    case Ace = 1
    case Two, Three, Four, Five, Six, Seven, Eight, Nine, Ten
    case Jack, Queen, King
    func simpleDescription() -> String {
        switch self {
```

```swift
        case .Ace:
            return "ace"
        case .Jack:
            return "jack"
        case .Queen:
            return "queen"
        case .King:
            return "king"
        default:
            return String(self.rawValue)
        }
    }
}
let ace = Rank.Ace
let aceRawValue = ace.rawValue
```

练习:写一个函数,通过比较它们的原始值来比较两个 Rank 值。

默认情况下,Swift 按照从 0 开始每次加 1 的方式为原始值进行赋值,不过可以通过显式赋值进行改变。在上面的例子中,Ace 被显式赋值为 1,并且剩下的原始值会按照顺序赋值。也可以使用字符串或者浮点数作为枚举的原始值。使用 rawValue 属性来访问一个枚举成员的原始值。

使用 init?(rawValue:)初始化构造器在原始值和枚举值之间进行转换。例如:

```swift
if let convertedRank = Rank(rawValue:3){
    let threeDescription = convertedRank.simpleDescription()
}
```

枚举的成员值是实际值,并不是原始值的另一种表达方法。实际上,如果没有比较有意义的原始值,就不需要提供原始值。例如:

```swift
enum Suit {
    case Spades, Hearts, Diamonds, Clubs
    func simpleDescription() -> String {
        switch self {
        case .Spades:
            return "spades"
        case .Hearts:
            return "hearts"
        case .Diamonds:
            return "diamonds"
        case .Clubs:
            return "clubs"
        }
    }
}
let hearts = Suit.Hearts
let heartsDescription = hearts.simpleDescription()
```

练习:给 Suit 添加一个 color()方法,对 Spades 和 Clubs 返回 black,对 Hearts 和 Diamonds 返回 red。

注意，有两种方式可以引用 Hearts 成员：给 hearts 常量赋值时，枚举成员 Suit.Hearts 需要用全名来引用，因为常量没有显式指定类型；在 switch 里，枚举成员使用缩写 .Hearts 来引用，因为 self 的值已经知道是一个 Suit。已知变量类型的情况下可以使用缩写。

使用 struct 来创建一个结构体。结构体和类有很多相同的地方，例如方法和构造器。它们之间最大的一个区别就是结构体是传值，类是传引用。例如：

```
struct Card {
    var rank:Rank
    var suit:Suit
    func simpleDescription() -> String {
        return "The \(rank.simpleDescription()) of \(suit.simpleDescription())"
    }
}
let threeOfSpades = Card(rank:.Three, suit:.Spades)
let threeOfSpadesDescription = threeOfSpades.simpleDescription()
```

练习：给 Card 添加一个方法，创建一副完整的扑克牌，并把每张牌的 rank 和 suit 对应起来。

一个枚举成员的实例可以有实例值。相同枚举成员的实例可以有不同的值。创建实例的时候传入值即可。实例值和原始值是不同的：枚举成员的原始值对于所有实例都是相同的，而且你在定义枚举的时候设置原始值。

例如，考虑从服务器获取日出和日落的时间，服务器会返回正常结果或者错误信息。例如：

```
enum ServerResponse {
    case Result(String, String)
    case Failure(String)
}
let success = ServerResponse.Result("6:00 am", "8:09 pm")
let failure = ServerResponse.Failure("Out of cheese.")
switch success {
case let .Result(sunrise, sunset):
    let serverResponse = "Sunrise is at \(sunrise) and sunset is at \(sunset)."
case let .Failure(message):
    print("Failure...  \(message)")
}
```

练习：给 ServerResponse 和 switch 添加第三种情况。

注意日升和日落时间是如何从 ServerResponse 中提取到并与 switch 的 case 相匹配的。

1.5.6 协议和扩展

使用 protocol 来声明一个协议。例如：

```
protocol ExampleProtocol {
    var simpleDescription:String { get }
    mutating func adjust()
}
```

类、枚举和结构体都可以实现协议。例如：

```
class SimpleClass:ExampleProtocol {
    var simpleDescription:String = "A very simple class. "
    var anotherProperty:Int = 69105
    func adjust(){
        simpleDescription += "  Now 100% adjusted. "
    }
}
var a = SimpleClass()
a.adjust()
let aDescription = a.simpleDescription

struct SimpleStructure:ExampleProtocol {
    var simpleDescription:String = "A simple structure"
    mutating func adjust(){
        simpleDescription += "(adjusted)"
    }
}
var b = SimpleStructure()
b.adjust()
let bDescription = b.simpleDescription
```

练习：写一个实现这个协议的枚举。

注意：声明 SimpleStructure 的时候，mutating 关键字用来标记一个会修改结构体的方法。SimpleClass 的声明不需要标记任何方法，因为类中的方法通常可以修改类属性（类的性质）。

使用 extension 来为现有的类型添加功能，如新的方法和计算属性。可以使用扩展在别处修改定义，甚至是从外部库或者框架引入的一个类型，使得这个类型遵循某个协议。例如：

```
extension Int:ExampleProtocol {
    var simpleDescription:String {
        return "The number \(self)"
    }
    mutating func adjust(){
        self += 42
    }
}
print(7.simpleDescription)
```

练习：给 Double 类型写一个扩展，添加 absoluteValue 功能。

可以像使用其他命名类型一样使用协议名。例如，创建一个有不同类型但是都实现一个协议的对象集合。当处理类型是协议的值时，协议外定义的方法不可用。例如：

```
let protocolValue:ExampleProtocol = a
print(protocolValue.simpleDescription)
// print(protocolValue.anotherProperty)   // Uncomment to see the error
```

即使 protocolValue 变量运行时的类型是 simpleClass，编译器也会把它的类型当作 ExampleProtocol。这表示不能调用类在它实现的协议之外实现的方法或者属性。

1.5.7 错误处理

使用采用 ErrorType 协议的类型来表示错误。例如：

```
enum PrinterError:ErrorType {
    case OutOfPaper
    case NoToner
    case OnFire
}
```

使用 throw 来抛出一个错误并使用 throws 来表示一个可以抛出错误的函数。如果在函数中抛出一个错误，这个函数会立刻返回，并且调用该函数的代码会进行错误处理。

```
func sendToPrinter(printerName:String) throws -> String {
    if printerName == "Never Has Toner" {
        throw PrinterError.NoToner
    }
    return "Job sent"
}
```

有多种方式可以用来进行错误处理。一种方式是使用 do...catch。在 do 代码块中，使用 try 来标记可以抛出错误的代码。在 catch 代码块中，除非另外命名，否则错误会自动命名为 error。例如：

```
do {
    let printerResponse = try sendToPrinter("Bi Sheng")
    print(printerResponse)
} catch {
    print(error)
}
```

练习：将 printer name 改为 "Never Has Toner" 使 sendToPrinter(_:) 函数抛出错误。

可以使用多个 catch 块来处理特定的错误。参照 switch 中的 case 风格来写 catch。例如：

```
do {
    let printerResponse = try sendToPrinter("Gutenberg")
    print(printerResponse)
} catch PrinterError.OnFire {
    print("I'll just put this over here, with the rest of the fire.")
} catch let printerError as PrinterError {
    print("Printer error: \(printerError).")
} catch {
    print(error)
}
```

练习：在 do 代码块中添加抛出错误的代码。需要抛出哪种错误来使第一个 catch 块进行接收？怎样使第二个和第三个 catch 进行接收呢？

另一种处理错误的方式使用"try?"将结果转换为可选的。如果函数抛出错误，该错误会被抛弃并且结果为 nil。否则，结果会是一个包含函数返回值的可选值。例如：

```
let printerSuccess = try? sendToPrinter("Mergenthaler")
let printerFailure = try? sendToPrinter("Never Has Toner")
```

使用 defer 代码块来表示在函数返回前，函数中最后执行的代码。无论函数是否会抛出错误，这段代码都将执行。使用 defer，可以把函数调用之初就要执行的代码和函数调用结束时的扫尾代码写在一起，虽然这两者的执行时机截然不同。例如：

```
var fridgeIsOpen = false
let fridgeContent = ["milk", "eggs", "leftovers"]
func fridgeContains(itemName:String) -> Bool {
    fridgeIsOpen = true
    defer {
        fridgeIsOpen = false
    }
    let result = fridgeContent.contains(itemName)
    return result
}
fridgeContains("banana")
print(fridgeIsOpen)
```

第 2 章 编写第一个应用程序

例如,在计算机屏幕上输出"Hello,world"这行字符串的计算机程序,其中文意思是"世界,你好"。通过本章,将学会使用开发环境和最简单的编程来实现 iOS 应用程序。

2.1 Swift 开发与学习环境

苹果公司在硬件和自身的操作系统方面提供了最好的安全性,不过这样也限制了很多开发者的便利。一般来讲,开发苹果系列产品的软件,必须拥有相应的苹果产品。

如果开发苹果手机应用程序,需要准备以下设备:

(1)苹果计算机,必须是基于 Intel 的 Macintosh 计算机。

(2)iPhone 或 iPod Touch,主要用来测试编写好的程序,也可以使用 Xcode 中的模拟器运行调试。

(3)苹果计算机操作系统 macOS。

(4)Xcode 开发工具,可以从苹果公司的 AppStore 免费下载。

2.1.1 Mac 开发环境 Xcode

Xcode 是苹果公司自己开发的一款功能强大的集成开发环境 IDE,只能运行在 macOS 系统上。较新的 Xcode8.3.2 支持 Swift 语言 3.1.1 版本,可以开发 iOS 10.3.1 的应用程序。

Xcode 可以编写 C、C++、Objective-C 和 Java 代码,可以生成 macOS 支持的所有类型的执行代码,包括命令行工具、框架、插件、内核扩展、程序包和应用程序。Xcode 具有编辑代码、编译代码、调试代码、打包程序、可视化编程、性能分析、版本管理等开发过程中所有的功能，而且还支持各种插件进行功能扩展、具有丰富的快捷键，可有效帮助开发人员提高效率。

Xcode 界面主要可以分成 5 个区域：工具栏、编辑区域、导航区域、调试和输出区域及功能区域，如图 2-1 所示。

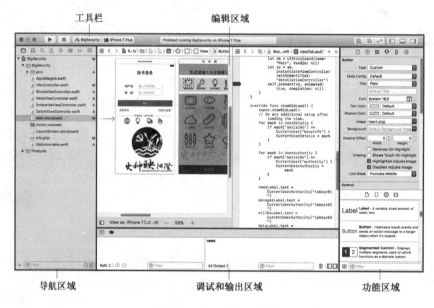

图 2-1　Xcode 界面功能

（1）工具栏主要包括快速运行程序按钮、选择模拟器或者真机按钮、标准编辑模式/辅助编辑模式/版本编辑模式、显示/隐藏左边导航区域、显示/隐藏底部调试输出区域、显示/隐藏右侧功能区域等。

（2）编辑区域主要包括 Storyboard 编辑/源代码编辑、断点设置等。

（3）导航区域主要包括工程的文件列表、底部文件过滤器，还有各种导航按钮：符号导航按钮、搜索导航按钮、问题导航按钮、测试导航按钮、调试导航按钮、断点导航按钮和报告导航按钮等。

（4）调试和输出区域：主要是断点中断或者程序出错后，在调试区域可以显示相关的变量等内容；输出区域主要是显示程序中 print 语句输出的内容。

（5）功能区域只要包括：文件查看器、帮助查看器、标识查看器、属性查看器、尺寸查看器、链接查看器，以及文件模板库、代码片段库、对象库、媒体库等。

2.1.2　Linux 下的学习环境

Swift 是苹果推出的最新编程语言，其目的是为了取代 Objective-C 成为构建 Mac OS X 和 iOS 的应用程序的主要语言。Swift 非常简洁易学，而且与其他流行的编程语言有类似的语法，容易上手。

然而，Swift 语言仅适用于苹果设备，所开发的程序不能在 Windows、Linux 或其他系统上运行。随着 Swift 日益普及,越来越多的开发人员都希望苹果将 Swift 开源,至少要让其能够支持更多的平台。苹果在听取了广大开发者的声音后,最终将 Swift 开源了,这意味着任何人都将可以在任何系统上使用。

现在,Linux 下已经可以部署 Swift 语言用于学习和开发 Web 端程序。目前,苹果只发布了针对 Ubuntu 的安装程序。

下面首先介绍一下在 Ubuntu 下如何安装部署 Swift。

(1)访问官方开源网站,下载相应文件。

在浏览器中输入 http://www.swift.org,找到 DOWNLOAD 选项,下载相应的文件,如图 2-2 所示。

图 2-2　访问 www.swift.org 下载相应文件

(2)安装 Swift,然后运行测试,如图 2-3 所示。

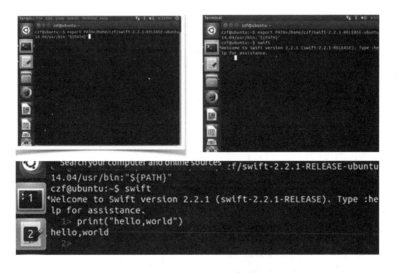

图 2-3　Swift 在 Ubuntu 中的运行

(3) 练习 Swift 语言，如图 2-4 所示。

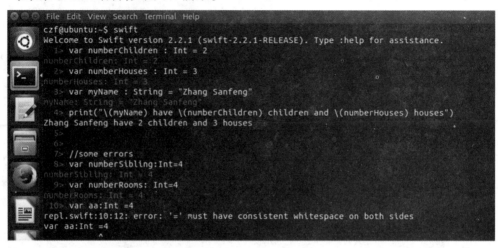

图 2-4　练习 Swift 语言

2.2　第一个 iOS 应用程序

这次通过 Xcode 主要实现两种不同效果的"Hello, world"。一种方法是通过 Xcode 中的工程项目来实现；另一种方法就是通过 Xcode 中的 Playground 来实现。

2.2.1　创建项目

1. 启动 Xcode

在 macOS 中找到 Xcode，然后运行。启动 Xcode 的常见方法有：从 Launch Pad 中找到 Xcode 图标；在 Spotlight 搜索中输入 Xcode，通过查找方式，找到 Xcode，双击然后即可运行；从 Dock 中运行 Xcode，如图 2-5 所示。

图 2-5　从 Dock 中运行 Xcode

2. 选择新建工程（Create a new Xcode project）

Xcode 的欢迎界面有 3 个选项：新建一个 playground 程序（Get started with a play ground）、新建一个工程 Xcode（Create a new Xcode project）和检查一个存在的工程（Check out an existing project），如图 2-6 所示。此处选择新建一个 Xcode 工程，即第 2 项。从欢迎界面，还可以了解到当前 Xcode 的版本为 8.3.2。

编写第一个应用程序 第2章

图 2-6　新建工程

3. 选择工程模板（Choose a template）

工程模板可以按照 iOS、watchOS、tvOS、macOS 和 Cross-platform 等类别进行选择。现在要开发 iOS 应用程序，则选择 iOS 大类，然后选择 Single View Application，如图 2-7 所示。

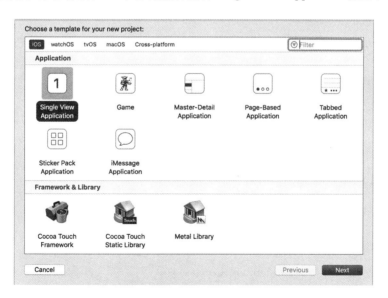

图 2-7　选择 iOS 类别的 Single View Application

4. 设置工程参数（Choose options）

工程参数主要包括：工程的名称（Product Name），填写 HelloWorld；团队（Team），选择 None；机构名称（Organization Name），填写 Zhifeng Chen，一般就是开发者的姓名；机构标识

（Organization Identifier），填写 cn.edu.szjm，一般就是开发者的域名；程序标识（Bundle Identifier），不需要填写，自动生成；程序语言（Language），选择 Swift；设备（Devices），有3个选项（Universal、iPad 和 iPhone），一般选择 iPhone；最后，还有3个参数，可以根据实际情况勾选，一般可以不选：是否使用 Core Data、是否包含单元测试（Unit Tests）和是否包含界面测试（UI Tests）等，如图2-8所示。

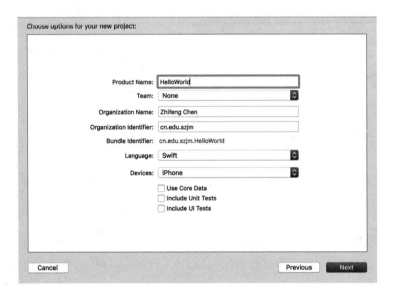

图2-8　设置工程参数

5. 选择存放的目录路径

一般从左边菜单选择保存的位置在桌面（Desktop），然后单击生成（Create）按钮，如图2-9所示。

图2-9　选择项目保存的目录路径

6. 设置完成，出现工程概要

工程各项参数设置完成，显示 Xcode 界面，中间部分为当前工程的概要情况，内容与

Info. plist 中相同,如图 2-10 所示。

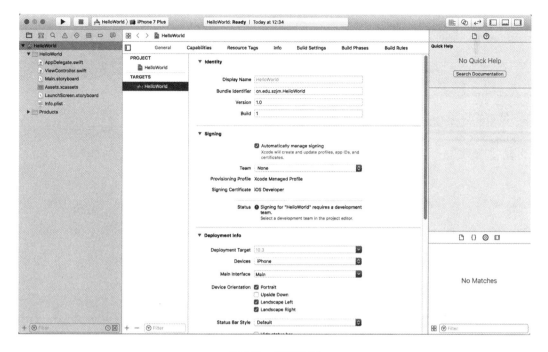

图 2-10　工程概要

7. 选择 ViewController. swift 文件,编写代码

一般从左边选择 ViewController. swift 文件,单击(如果用户不小心双击,则会弹出一个窗口,可以关闭后,重新单击),中间的编辑区域会显示这个文件的源代码。

现在用户可以在 override func viewDidLoad()函数的相应位置,输入相应的语句 print("Hello,world"),如图 2-11 所示。

图 2-11　选择 ViewController. swift,编辑区域显示源代码

8. 运行并查看输出区域结果

单击工具栏上的运行按钮![图标]，系统会自动弹出一个空白模拟器，同时在输出区域会出现一行文字"Hello,world"，如图2-12所示。

图2-12　运行查看输出区域结果

注意：如果显示分辨率设置不够，手机模拟器会显得很大，可以选择手机模拟器，然后选择Window→Scale→50%命令，如图2-13所示。

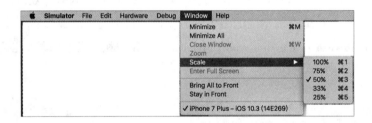

图2-13　模拟器显示比例设置

9. 选择Main.storyboard中View Controller界面进行设计

一般从左边菜单选择Main.storyboard文件，单击，在编辑区域会出现手机可视化设计界面。现在里面只有一个View Controller界面，从右边功能区域选择一个Label控件，拖放到这个View Controller界面中，如图2-14所示。

10. 拖放Label到View Controller中

将Label控件拖放到手机界面的合适位置，单击Label，可以修改其内容，如图2-15所示。

11. 修改Label内容

现在可以将Lable的内容修改为"Hello,world"，同时可以根据文字内容的多少，调整

Label 的大小，如图 2-16 所示。

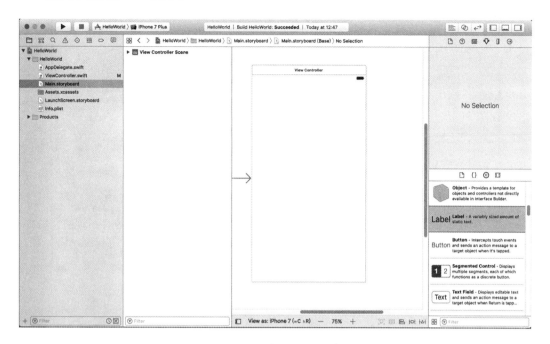

图 2-14　可视化界面设计

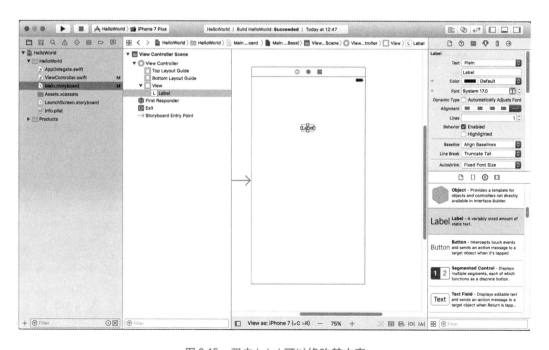

图 2-15　双击 Label 可以修改其内容

12. 运行并查看模拟器结果

单击 ▶ 运行工程，可以看到输出区域和手机模拟器的结果，如图 2-17 所示。

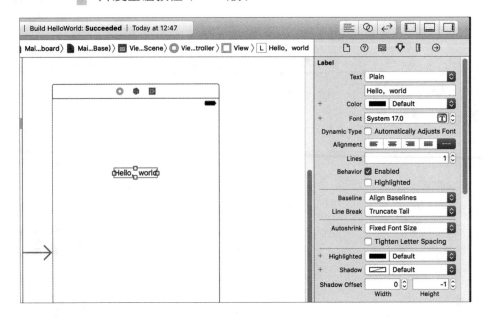

图 2-16　Label 内容修改为"Hello,world"

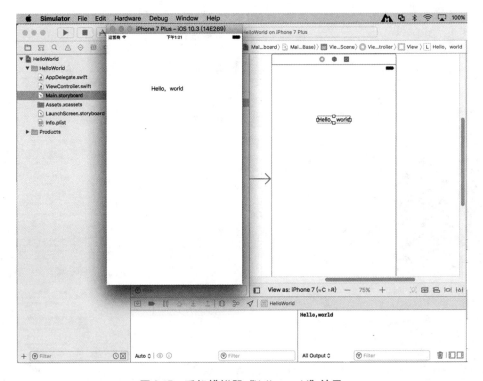

图 2-17　手机模拟器"Hello,world"结果

2.2.2　常见 UI 组件

Xcode 开发环境提供了各种各样的 UI 组件(见图 2-18),从而大大减轻了开发人员的负担。

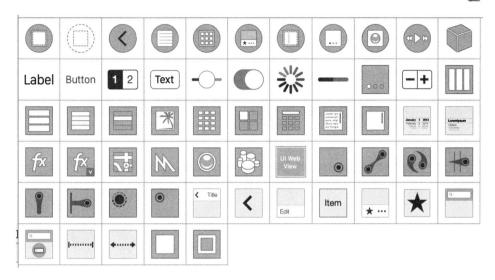

图 2-18　UI 组件

1. UILabel

UILabel 常用于在界面上显示一行或多行文本,可视化创建的方法是将这个组件拖放到 Main.storyboard 中某一个 View Controller 中。

注:具体可参考图 2-15 右侧的功能区域,下同。

UILabel 的代码创建方法:

```
//方法1:创建时设置 frame
let rect = CGRect(x:10, y:10, width:300, height:30)
let label = UILabel(frame:rect)
//添加到 self.view 上才会显示出来
self.view.addSubview(label)

//方法2:先创建,后设置 frame
let label = UILabel()
label.frame = CGRect(x:10, y:90, width:300, height:30)
self.view.addSubview(label)
```

UILabel 的常用属性:文本内容设置、背景色设置、字体颜色设置、字体大小设置和文本对齐方式设置等。例如:

```
//设置背景色
label.backgroundColor = UIColor.green
//设置字体颜色
label.textColor = UIColor.red
```

```
//设置字体大小
label.font = UIFont.systemFont(ofSize:14)
//设置文本对齐方式,默认左对齐
label.textAlignment = NSTextAlignment.right
//设置要显示的文本
label.text = "I am a label"
//当文字超出标签宽度时,自动调整文字大小,使其不被截断
label.adjustsFontSizeToFitWidth = true
```

显示多行文本:显示多行文本需要两个条件:一个是设置 numberOfLines 的值为要显示的行数;二是 label 的高度要大于或等于文字的行高和文字行间距总和。

```
//方法1:显示固定行数的文本
label.frame = CGRect(x:10, y:100, width:100, height:100)
label.numberOfLines = 2
label.text = "I am a label,I am a label,I am a label"

//方法2:根据文字长度确定显示几行,只需设置 numberOfLines = 0
label.numberOfLines = 0
label.text = "I am a label,I am a label,I am a label"
```

练习:请设计实现新闻界面,采用多个 Label,要求如图 2-19 所示。

图 2-19　多个 UILabel 设计出新闻界面

2. UIButton

UIButton 是一个用于接受用户触摸事件的 iOS 常用组件,可视化创建的方法是将这个

组件拖放到 Main.storyboard 中某一个 View Controller 中。UIButton 常用的触摸事件类型：

（1）touchDown：单点触摸按下事件，点触屏幕。

（2）touchDownRepeat：多点触摸按下事件，点触计数大于 1，按下第 2、3 或第 4 根手指的事件。

（3）touchDragInside：触摸在控件内拖动事件。

（4）touchDragOutside：触摸在控件外拖动事件。

（5）touchDragEnter：触摸从控件之外拖动到内部事件。

（6）touchDragExit：触摸从控件内部拖动到外部事件。

（7）touchUpInside：在控件之内触摸并抬起事件。

（8）touchUpOutside：在控件之外触摸抬起事件。

（9）touchCancel：触摸取消事件，即一次触摸因为放上太多手指而被取消，或者电话打断。

Button - Intercepts touch events and sends an action message to a target object when it's tapped.

UIButton 的代码创建方法：

```
//方法1:创建一个系统内建样式的UIButton
let button = UIButton(type:UIButtonType.system)
button.frame = CGRect(x:150, y:150, width:120, height:40)
button.setTitle("Click me", for:UIControlState.normal)
button.titleLabel?.font = UIFont.systemFont(ofSize:18)
button.addTarget(self, action:#selector(btnClick(_:)), for:UIControlEvents.touchUpInside)
button.layer.cornerRadius = 5.0
//添加到self.view上才会显示出来
self.view.addSubview(button)

//方法2:创建一个自定义的UIButton,用于模拟复选框
let rect = CGRect(x:10, y:150, width:50, height:50)
let button = UIButton(frame:rect)
button.setImage(UIImage(named:"checknoneBtn.png"), for:UIControlState.normal)
button.setImage(UIImage(named:"checkedBtn.png"), for:UIControlState.selected)
button.addTarget(self, action:#selector(btnClick(_:)), for:UIControlEvents.touchUpInside)
button.layer.cornerRadius = 5.0
self.view.addSubview(button)

//Button触摸后,需要调用的事件方法
func btnClick(_ sender:UIButton){
    print("Wuwu ~ ~, I am here now")
    sender.isSelected = !sender.isSelected
}
```

可以模拟复选框的 UIButton，需要两个状态的图片（见图 2-20），左边是选中后显示的图

片(checkedBtn. png),右边是正常状态下的图片(checknoneBtn. png)。

图 2-20　模拟复选框的 UIButton 所需要的两张图片

UIButton 的常用属性和方法:按钮上的文字设置 setTitle()、按钮上的文字颜色设置 setTitleColor()、按钮上的文字字体大小设置 titleLabel?. font、按钮背景颜色设置 backgroundColor 和是否可用 isEnable、是否隐藏不显示 isHidden、是否被选中 isSelected 等。

练习:请设计实现调查界面,采用 UIButton 和 UIlabel 设计调查选项内容,要求如图 2-21 所示。

3. UIImageView

UIImageView 是一个用于 iOS 显示图片、简单动画等的组件,可视化创建的方法是将这个组件拖放到 Main. storyboard 中某一个 View Controller 中。UIImageView 的用户交互默认是关闭的,也就是说 ImageView 对其上的触摸事件都不会响应,通过设置 userInteractionEnabled 属性为真,然后给 UIImageVIew 添加一个点击手势,就可以实现对触摸事件的响应。

图 2-21　调查选项界面

 Image View - Displays a single image, or an animation described by an array of images.

UIImageView 的代码创建方法:

```
// 创建一个 UIImageView,图片文件为 checkedBtn. png
let imageViewRect = CGRect(x:0, y:350, width:100, height:100)
let imageView = UIImageView(frame:imageViewRect)
let img = UIImage(named:"checkedBtn. png")
imageView. image = img
//添加到 self. view 上才会显示出来
self. view. addSubview(imageView)
```

UIImageView 中提供了存储多张图片来创建动画的功能,具体做法是在 animationImages 属性中设置一个图片数组,然后使用 startAnimating 方法开始动画,最后用 stopAnimating 方法停止动画。同时,使用 animationDuration 属性可以设置动画每帧切换的速度(秒)。

UIImageView 的交互功能代码创建方法:

```
//用户交互
imageView.isUserInteractionEnabled = true
let tap = UITapGestureRecognizer(target:self, action:#selector(tapAction(tap:)))
imageView.addGestureRecognizer(tap)

//UIImageView 触摸后,需要调用的事件方法
func tapAction(tap:UITapGestureRecognizer){
    let scale:CGFloat = 1.2
    var frame = tap.view!.frame
    frame = CGRect(x:frame.origin.x , y:frame.origin.y , width:frame.size.width*scale, height:frame.size.height*scale)
    tap.view!.frame = frame
}
```

4. UIView

UIView 表示屏幕上一块矩形区域,它在 APP 中占有绝对重要的地位,因为 iOS 中几乎所有的可视控件都是 UIView 的子类,可视化创建的方法是将这个组件拖放到 Main.storyboard 中某一个 View Controller 中。

UIView 继承自 UIResponder,它是负责显示的画布,如果说把 window 比作画框,我们就是不断地在画框上移除、更换或者叠加画布,或者在画布上叠加其他画布,大小由绘画者来决定。有了画布,就可以在上面进行绘画。UIView 的功能:管理视图区域里的内容、处理视图区域中的事件、管理子视图,以及绘图、动画等。

UIView 的常用属性:

(1) frame:相对父视图的坐标和大小(x,y,width,height)。

(2) bounds:相对自身的坐标和大小,所以 bounds 的 x 和 y 永远为 0(0,0,width,height)。

(3) center:相对父视图中心位置的坐标。

(4) transform:控制视图的放大、缩小和旋转。

(5) superview:获取父视图。

(6) subviews:获取所有子视图。

(7) alpha:视图的透明度(0.0-1.0)。

(8) tag:视图的标志(Int 类型,默认等于 0),设置后,可以通过 viewWithTag() 方法拿到这个视图。

（9）backgroundColor：背景颜色。

UIView 的常用方法：

（1）func addSubview(view:UIView)：添加视图到父视图，只要越晚添加，视图就在越上层，类似于画图软件中的图层概念。

（2）func removeFromSuperview()：将视图从父视图中移除。

（3）func exchangeSubview(at:index1, withSubviewAt:index2)：将 index1 和 index2 位置的两个视图互换位置。

（4）func bringSubview(toFront:UIView)：把视图移到最顶层。

（5）func sendSubview(toBack:UIView)：把视图移到最底层。

（6）func viewWithTag(tag:Int) -> UIView?：根据 tag 值获取视图。

UIView 的代码创建方法：

```
//创建 View
let view1 = UIView()
let view2 = UIView(frame:CGRect(x:20,y:120, width:100,height:100))
let view3 = UIView(frame:CGRect(x:40,y:140, width:100,height:100))
//设置 view 的尺寸
view1.frame = CGRect(x:0,y:100, width:100,height:100)
//设置 view 的背景色
view1.backgroundColor = UIColor.red
view2.backgroundColor = UIColor.green
view3.backgroundColor = UIColor.blue
//设置 view 的中心位置,不改变 view 的大小
view1.center = CGPoint(x:80,y:200)
//依次添加三个视图(从上到下是:蓝,绿,红)
self.view.addSubview(view1)
self.view.addSubview(view2)
self.view.addSubview(view3)
//把 view1(红)移到最上面
self.view.bringSubview(toFront:view1)
//设置 view 的透明度
view1.alpha = 0.5
//设置 view1 的圆角角度
view1.layer.cornerRadius = 10
//设置边框的宽度
view1.layer.borderWidth = 2
//设置边框的颜色
view1.layer.borderColor = UIColor.red.cgColor
```

在实际使用中，可以将 UIView 作为一个容器，把其他的组件放在这个 UIView 上面，这样可以形成一个整体，便于整体移动等处理。

2.2.3 交互设计

在开发 iOS 应用程序时，一般可以通过在 Main.storyboard 中把项目的 UI 界面和一些页面关系实现起来，这样可以更容易把握所开发的程序，实现更好的人机交互。

1. 显示原理

首先从过去的 CRT 显示器原理说起。CRT 的电子枪按照上面方式,从上到下一行行扫描,扫描完成后显示器就呈现一帧画面,随后电子枪回到初始位置继续下一次扫描。为了把显示器的显示过程和系统的视频控制器进行同步,显示器(或者其他硬件)会用硬件时钟产生一系列定时信号。当电子枪换到新的一行,准备进行扫描时,显示器会发出一个水平同步信号(Horizonal Synchronization),简称 HSync;而当一帧画面绘制完成后,电子枪回复到原位,准备画下一帧前,显示器会发出一个垂直同步信号(Vertical Synchronization),简称 VSync。显示器通常以固定频率进行刷新,这个刷新率就是 VSync 信号产生的频率,如图 2-22 所示。

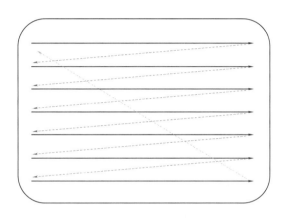

图 2-22 扫描与显示原理

现在的设备普遍采用了 LED 液晶显示屏,但基本原理仍然没有变。

通常来说,计算机系统中 CPU、GPU、显示器是以上面这种方式协同工作的。CPU 计算好显示内容提交到 GPU,GPU 渲染完成后将渲染结果放入帧缓冲区,随后视频控制器会按照 VSync 信号逐行读取帧缓冲区的数据,经过可能的数模转换传递给显示器显示,如图 2-23 所示。

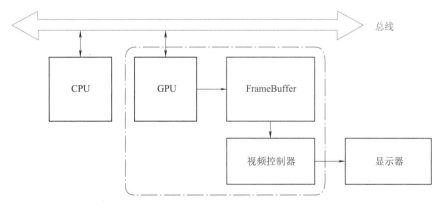

图 2-23 GPU 显示原理

在最简单的情况下,帧缓冲区只有一个,这时帧缓冲区的读取和刷新都会有比较大的效率问题。为了解决效率问题,显示系统通常会引入两个缓冲区,即双缓冲机制。在这种情况下,GPU 会预先渲染好一帧放入一个缓冲区内,让视频控制器读取,当下一帧渲染好后,GPU 会直接把视频控制器的指针指向第二个缓冲器,如此一来效率会有很大的提升。

双缓冲虽然能解决效率问题,但会引入一个新的问题。当视频控制器还未读取完成时,即屏幕内容刚显示一半时,GPU 将新的一帧内容提交到帧缓冲区并把两个缓冲区进行交换后,视频控制器就会把新的一帧数据的下半段显示到屏幕上,造成画面撕裂现象。

为了解决这个问题,GPU 通常有一个机制叫作垂直同步(简写也是 VSync),当开启垂直同步后,GPU 会等待显示器的 VSync 信号发出后,才进行新的一帧渲染和缓冲区更新。这样能解决画面撕裂现象,也增加了画面流畅度,但需要消费更多的计算资源,也会带来部分延迟。

2. 布局计算

视图布局的计算是 APP 中最常见的消耗 CPU 资源的地方。如果能在后台线程提前计算好视图布局,并且对视图布局进行缓存,那么这个地方基本就不会产生性能问题。

不论通过何种技术对视图进行布局,其最终都会落到对 UIView.frame/bounds/center 等属性的调整上。对这些属性的调整非常消耗资源,所以尽量提前计算好布局,在需要时一次性调整好对应属性,而不要多次、频繁地计算和调整这些属性。

Autolayout 是苹果本身提倡的技术,在大部分情况下也能很好地提升开发效率,但是 Autolayout 对于复杂视图来说常常会产生严重的性能问题。随着视图数量的增长,Autolayout 带来的 CPU 消耗会呈指数级上升。

如果不想手动调整 frame 等属性,可以用一些工具方法替代(比如常见的 left/right/top/bottom/width/height 快捷属性),或者使用 ComponentKit、AsyncDisplayKit 等框架。

如果一个界面中包含大量文本(比如微博、微信、朋友圈等),文本的宽高计算会占用很大一部分资源,并且不可避免。如果对文本显示没有特殊要求,可以参考 UILabel 内部的实现方式:用 boundingRectWithSize 来计算文本宽高,用 drawWithRect:options 来绘制文本。尽管这两种方法性能不错,但仍旧需要放到后台线程进行以避免阻塞主线程。

如果用 CoreText 绘制文本,就可以先生成 CoreText 排版对象,然后自己计算,并且 CoreText 对象还能保留以供稍后绘制使用。

屏幕上能看到的所有文本内容控件,包括 UIWebView,在底层都是通过 CoreText 排版、绘制为 Bitmap 显示的。常见的文本控件(UILabel、UITextView 等),其排版和绘制都是在主线程进行的,当显示大量文本时,CPU 的压力会非常大。对此解决方案只有一个,那就是自定义文本控件,用 TextKit 或最底层的 CoreText 对文本进行异步绘制。尽管这实现起来非常麻烦,但其带来的优势也非常大,CoreText 对象创建好后,能直接获取文本的宽高等信息,避免了多次计算(调整 UILabel 大小时算一遍、UILabel 绘制时内部再算一遍);CoreText 对象占用内存较少,可以缓存下来以备稍后多次渲染。

当用 UIImage 或 CGImageSource 的那几个方法创建图片时,图片数据并不会立刻解码。图片设置到 UIImageView 或者 CALayer.contents 中,并且 CALayer 被提交到 GPU 前,CGImage 中的数据才会得到解码。这一步是发生在主线程的,并且不可避免。如果想要绕开这个机制,常见的做法是在后台线程先把图片绘制到 CGBitmapContext 中,然后从 Bitmap 直接创建图片。目前常见的网络图片库都自带这个功能。

图像的绘制通常是指用那些以 CG 开头的方法把图像绘制到画布中,然后从画布创建图片并显示这样一个过程。最常见的地方就是 UIView 的 drawRect 中。由于 CoreGraphic 方法通常都是线程安全的,所以图像的绘制可以很容易地放到后台线程进行。

3. 用户界面

(1)"明确"应该放在设计的首要位置。对任何界面而言,"明确"是首要的也是最重要的一点。人们必须能够辨别出它是什么,才能有效地使用设计出来的界面。设计师在设计的时候,要去关心人们为何会使用这个应用,去了解什么样的界面是能帮助他们与之互动的,去预测人们在使用时的行为并能够成功地反馈给他们。这样做了之后,界面中再出现需要推理的地方以及延时反应都是可以被容忍的,但是绝对不能出现让用户困惑的地方。明确的界面能够给使用者进一步操作的信心。一个应用就算有一百张页面,但是每一页都是清晰明确的,它也远胜于只有一页却不知所云的应用。图 2-24 所示展示了内容的主次有序编排。

图 2-24　内容的主次有序编排

(2)界面是为了交互而存在。界面的存在是为了让人和我们的世界产生互动。它可以帮助人们理清、明白、使用、展示相互之间的关系,它能够把人们聚集在一起,也可以将人们隔开,实现人们的价值并为人们服务。界面设计不是艺术设计,也不是用来标榜设计师的个人。界面的功用和效果是可以被测量的,但是它们不是功利性的。优秀的界面不但能够让人们做事有效率,还能够激发、唤起和加强人们与这个世界的联系。图2-25所示的黑白色调有利于快速选择交换。

图2-25 黑白色调有利于快速选择交互

(3)不惜一切代价吸引用户注意。人们生活在一个容易被打扰的世界,很难在一个不被干扰的环境中静下心来阅读。用户注意力是非常宝贵的,所以,不要在应用的周围放一些容易令人分心的东西。

要把设计这个画面最初的目的时刻放在首位。如果用户正在阅读,那先让他们专心地读完之后再弹出广告(如果一定要放广告的话)。尊重用户的注意力,不仅会让用户感到高兴,本身的设计也会收获好结果。如果在界面设计中,用户使用是首要目标,那么尊重用户的注意力是先决条件,要不惜一切代价保护它。图2-26所示为用颜色吸引用户眼球。

(4)让用户掌控一切。人们会在自己能掌控的环境中感觉最舒心,最放松。设计草率的软件应用不但剥夺了这种舒适性,还会迫使人们面对毫无预期的互动、困惑的流程和意外的结果。通过定期地梳理系统状态,描述因果关系(如果做了,就会被体现出来),并且在每一步操作都给出提示,让用户感觉每一步操作都在他的掌控中。

(5)直观操作是最好的。好的界面是无意识的,就像人们在实际生活中直接操作一样。这并不是容易实现的,并且随着元素和资讯的不断增加,这就变得更难,所以需要对界面进行精心的设计。要想在页面上添加一个不必要的东西非常简单,例如,加个华丽的按钮、镶边、图形、选项、偏好、窗口、附件和其他一些冗余的东西,以至于人们一头扎进处理界面元素细节的怪圈中而忽视了最重要的事情。最重要的,应该抓住直观操作这个最初的目标。界面设计要尽可能简洁。图2-27所示为具有一致性操作的界面。

编写第一个应用程序 第 2 章

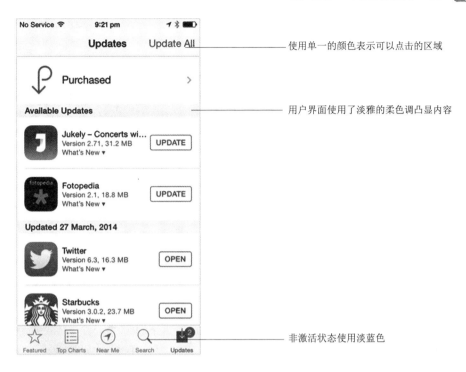

图 2-26　颜色是吸引用户眼球的不二法门

图 2-27　一致性有助于操作的便利

（6）一个页面一个主要操作。我们设计的每一个画面,都应给用户提供有实际意义的单一操作。这一点,令界面易于操作,如果有需要,新增或扩充也更简易。如果一个页面上有两个或两个以上的主要操作,瞬间就会让用户感到困惑。就像写文章要有单一的、强有力的论点一样,界面设计中的单个页面,也都应该有单一且明确的操作,这是它存在的理由。

(7)让次要操作在次要位置。页面在包含一个主要动作的同时,可以有多个次要动作,但尽量不要让它们喧宾夺主。就像写一篇文章的目的,是为了让人可以阅读、可以了解,而不是为了人们能够把它转载到社交网络上。让次要动作放在次要的位置,削减它们的视觉冲击力,或者在主要动作完成后再显示它们。

(8)提供自然而然的下一步操作。很少有交互是故意被放在最后的,所以要为用户精心设计交互的下一步操作。要预期用户下一步的交互是怎样的,并且通过设计将其实现。当用户已经完成要做的操作后,别让他们不知所措地停留在那里,提供自然而然的下一步,帮助他们完成操作。

(9)界面外观遵循用户行为。人总是对符合期望的行为最感舒适。当某人或某件事的行为始终按照人们所期望的那样去进行时,人们会感觉到和他们之间的关系不错。因此,设计出来的元素看起来应该像它们本身特征一样。在具体操作中,这意味着用户只要看到这个界面元素,就应该能猜测出这个元素是做什么的。如果看起来是个按钮,它就应该有按钮的功能,不要在基本的交互问题上耍小聪明,请把创造力留到更高层次的需求上。图2-28所示为简单清晰的页面。

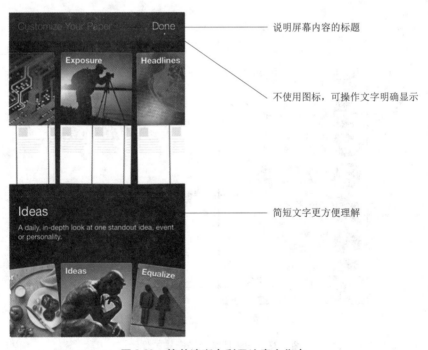

图2-28　简单清晰有利于注意力集中

(10)前后一致的重要性。遵循上一规则,画面中视觉元素的外观不应该是一样的,除非它们的功能相近。如果是功能相同或相近的元素,那么它们的外观就应该是类似的;反之,如果元素各自的功能不同,那么它们的外观也应该不同。

(11)强烈的视觉层次会让效果更好。强烈的视觉层次会让画面有清晰的浏览次序。也就是说,当用户每次都用相同的顺序浏览同样的东西,微小的视觉层次令使用者不知道哪里才是需要注意的重点,最后只会让用户感到困惑和混乱。在不断变化的设计环境中,保持强烈的视觉层次是很困难的,因为所有元素视觉上的重量是相对的:当所有文字都是粗体,那就没有所谓的"粗体"了。如果要在画面中添加一个视觉强烈的元素,设计者应该重新调整页面上所有元素的重量分配,来达到强烈视觉层次的效果。大多数人都不会注意到视觉重量这一点,但它其实是强化(弱化)设计的最简单的方法。

(12)巧妙的布局减轻用户认知负担。恰当地编排画面上的元素能够以少见多,帮助用户更加快速简单地理解设计的界面,因为已经用设计清楚地说明了它们彼此之间的关系。用方位和方向上的编排可以自然地将相同的元素联系在一起。通过对内容的巧妙编排,可以减轻用户的认知负担,他们不再需要花时间去思考元素之间的关联。不要迫使用户去思考,要将你的设计直接呈现给用户。

(13)重点不要总是用颜色来强调。物体的色彩会随光线改变而改变。艳阳高照下与夕阳西下时,同一棵树也会呈现不同的景象。在自然世界中,色彩很容易受环境影响而改变,所以在设计时,色彩不应该占很大的比重。作为辅助,可以用高亮的颜色吸引人们注意力,但这不是用于强调的唯一方法。在长篇阅读或者长时间对着计算机屏幕的情况下,可以使用柔和的背景降低亮度。当然,也可用活泼亮丽的色彩当背景,但是要确保适合读者。

(14)逐步说明。只在画面中显示必要的信息。如果用户需要做出决定,只要展现足够的信息供其选择,然后他们会到下一页去寻找更多的细节。避免一次呈现或解释全部的信息,如果可以,将选择放在后面的画面展示,这会使设计的界面交互更加清晰。

(15)内置帮助。在理想的界面设计中,"帮助"选项没有必要,因为界面操作是有引导性的。"帮助"的下一步实际上是内嵌在上下文中的"帮助",只有在用户确实需要时才显示,平常应该是隐藏的状态。让用户自己去寻求帮助。重要的是要保证用户可以顺畅地使用所设计的界面。

(16)重要时刻:初始状态。第一次使用界面的体验是非常重要的,而这却常常被设计师忽略。为了让用户更快地上手,最好在设计时保持初始状态,也就是还没开始使用过的状态。这个状态不是一张空白的画布,应该要提供一个方向和指引,令用户迅速进入状况。在初始状态下的互动过程中可能会存在一些摩擦,一旦用户了解了规则,就会有很高的机会获得成功。

(17)好的设计是隐形的。好的设计有一个奇怪的特性,它通常很容易被用户忽略。其中的一个原因是。这个设计非常成功,以至于用户完全专注在他想要达到的目标,而不是这个界面。当用户顺利地完成目的时,他们会感到很满意,并不需要反映任何问题。作为一名设计师,这样会很困难,因为他们很少收到正面的回应,很少知道哪些设计是好的。但是优秀的设计师是满足于设计出好用的界面,并且他们知道满意的使用者通常是沉默的。图 2-29 展示了好的图标设计。

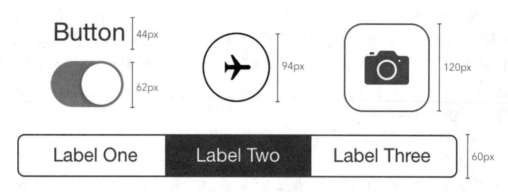

图 2-29　图标的设计是隐形的

（18）从其他设计领域下手。视觉、平面设计、排版、文案、信息架构和视觉设计……所有这些学科都是界面设计的一部分,它们都是可以被涉猎和研究的,不要企图跟它们划分界线。

（19）界面的存在是为了被使用。在大多数设计领域,有用户使用就说明界面设计得成功。就像是一个漂亮椅子,虽然漂亮,但坐起来不舒服,用户就不会选择它,它就是一个失败的设计。因此,界面的存在是为了尽可能多地创造好用的环境让用户使用。设计师设计作品如果仅仅是拿来满足自己的虚荣心,是远远不够的,它必须要被使用。

2.2.4　程序入口

iOS 其实和其他语言(Java、C)一样,也是通过一个 main() 函数作为入口,main() 函数封装在了 UIApplication 里面,系统会自动调用,这就是 iOS 应用程序的启动入口。

在启动过程中,UIApplication 会扫描 Info.plist,找到需要加载的入口 storyboard。如 Main.storyboard,读取里面的 UIViewController,然后就能显示相应的界面,响应相应的事件,执行代码,处理各种功能。iOS 启动流程及事件触发如图 2-30 所示。

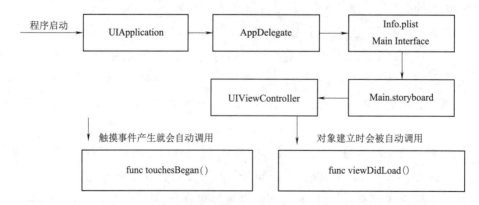

图 2-30　iOS 启动流程及事件触发

UIApplicationMain( )函数调用创建一个 UIApplication 对象及程序代理对象(通常为：AppDelegate)，UIApplication 对象扫描 Info. plist 文件，将其中 Main storyboard file base name 所指定的 Storyboard 文件装入(通常为：Main. storyboard)，如图 2-31 所示。

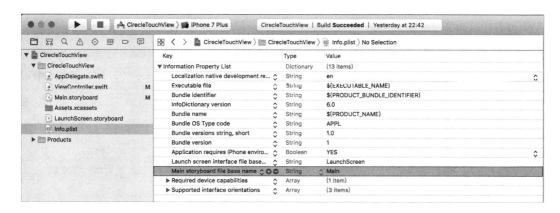

图 2-31　Info. plist 指定启动 Main. storyboard

UIApplication 对象从程序代理对象 AppDelegate 中获取窗口对象 UIWindow，也可以创建一个 UIWindow 新实例并将其与程序代理对象相关联。

将 Storyboard 文件中 Initial View Controller 属性所指定的 UIViewController 实例化，并将它赋予为 UIWindow 的 root View Controller，如图 2-32 所示。

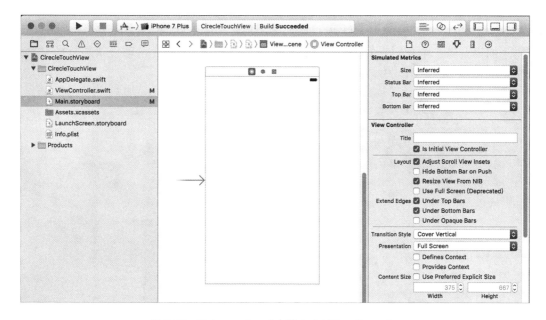

图 2-32　Main. storyboard 中的 Intial View Controller

最后向程序代理对象 AppDelegate 发送完成启动消息，以便程序员做自己的初始化工作。

2.3 Outlet 和 Action

在 iOS 的开发中，有两个重要的概念：Outlet 和 Action，它们可以很简单地把程序代码和界面组件、事件触发等连接起来。

(1) Outlet：是一种特别定义的变量。通过 Outlet，可以从与 Outlet 命名对应的控件中取出信息，或者将新的信息赋予控件。在 Main.storyboard 界面上配置的每个控件，都可以通过 Outlet 与代码连接，从而让程序员可以在程序中存取 Outlet 中的内容。在这里，Outlet 是将界面组件和代码变量连接的一种纽带，实现了界面组件与代码间的信息交换。

(2) Action：是一种特别定义的函数，是将触摸、晃动等行为事件与函数功能相连接。当相应的事件发生时，Action 所对应的函数内的代码将被自动执行。

所以，IBOutlet 和 IBAction 都只是一个标记，IBOutlet 用来标记代码片段中的变量，这个变量是应该和界面中某个对象相关联的；IBAction 应该用来标记代码片段中的方法(函数)，这个方法是应该和界面中的某个对象相关联的，用来响应对象应该响应的操作。

2.3.1 Outlet

Outlet 是一种特别定义的变量，在代码中对应的变量前面加上 IBOutlet 来修饰标记，本身来说它只是个标记。Outlet 要和 nib 文件(也就是 storyborad 中的界面)中的一个对象关联起来(建立了一个 connection)，一旦建立了连接，就可以在程序中进行赋值或者读取其内容。

下面以开发一个新闻界面(见图 2-33)为例，当程序运行后，会自动将新闻数据更新到界面上的相应 Outlet 中。

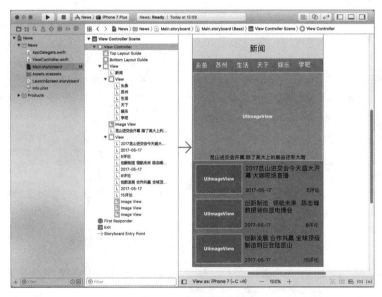

图 2-33　新闻客户端

1. 项目界面规划

根据新闻客户端的实际运行情况,界面设计如图 2-34 所示。

通过分析可以看到,界面上的各个元素可以分解为三大类:UIView、UILabel 和 UIImageView,具体如图 2-35 所示。

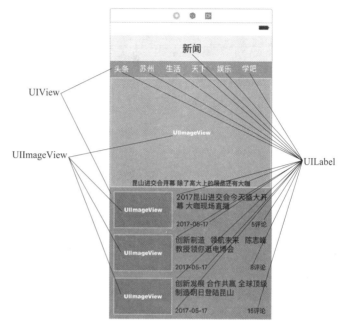

图 2-34 新闻客户端界面　　　　　　图 2-35 界面元素分解

2. 新建工程 News,保存到桌面

新建一个 Xcode 工程,名称为 News,保存到桌面上,和第一个 iOS 程序的步骤一致。

3. 打开 Main.storyboard,找到 View Controller

打开 Main.storyboard,选择 View Controller,单击文件查看器(File Inspector),找到两个选项:Use Auto Layout 和 Use Trait Variations,如图 2-36 所示。

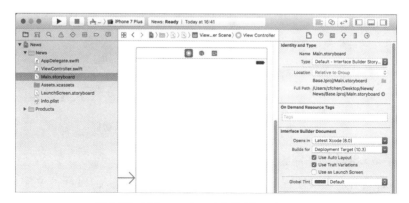

图 2-36 Main.storyboard 中的 View Controller

4. 取消 User Auto Layout 和 Use Trait Variations 功能

在取消这两个功能的时候，会出现一个提示，选择同意即可，如图 2-37 所示。

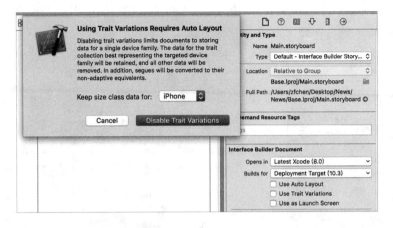

图 2-37　确认取消 Trait Variations 提示框

5. 拖放 Label 到界面中，调整字体大小和 Autoresizing 等

用户在对象库（Object Library）中找到 UILabel，将其拖放到界面的合适位置，进入尺寸查看器（Size Inspector），调整 UILabel 的大小（width 和 height），设置 Autoresizing 功能，如图 2-38 所示。

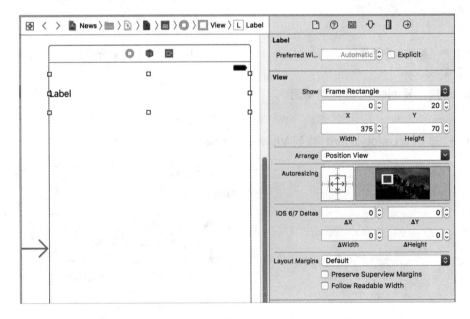

图 2-38　设置 Label 的大小和 Autoresizing 功能

6. 设置标题 Label 的字体和背景色

选择 Label，然后进入属性查看器（Attributes Inspector），输入 Label 的文字内容"新闻"，调整字体大小 Font，设置整个 Label 的背景色 Background，如图 2-39 所示。

图 2-39　Label 字体大小及背景色

7. 运行查看进展

模拟器根据需要可以进行向左旋转（Rotate Left）、重启（Reboot）等，如图 2-40 所示。

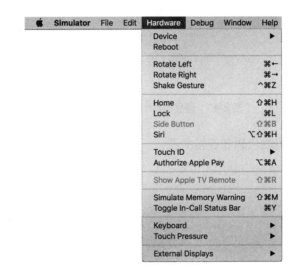

图 2-40　模拟器的 Hardware 菜单

单击工具栏上的"运行"按钮，并将模拟器进行向左旋转（选择 Hardware→Rotate Left 命令），将会出现如图 2-41 所示效果。

8. 建立一个 Outlet，用于连接代码和标题

在 Assistant Editor 情况下，会同时出现界面和代码。这样用户可以在界面区域选择 Label，然后按住【Ctrl】键不放，用鼠标拖到 ViewController 类的代码中，具体可以在 viewDidLoad( ) 函数上方空白位置，放开鼠标左键，弹出如图 2-42 所示对话框，输入名称 Name 为 titleLabel。若空白处不够，可以通过换行回车增加空白位置，这样可以让代码更清楚。

图 2-41　模拟器旋转后的效果

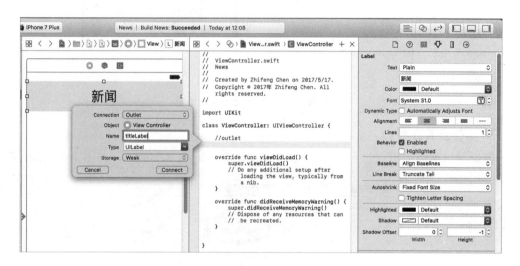

图 2-42　建立 Outlet 连接界面和代码

9. 查看 Outlet 相关的代码和连接关系

如图 2-43 所示，可以获知这个 Outlet 代表的是一个 Label，其变量名称为 titleLabel，变量类型为 UILabel。这个 Outlet 前面还有一个实心的圆圈，将鼠标指针放到这个圆圈上，相应界面的 Label 会变颜色，表示连接已经存在。

（1）@ IBOutlet 告知编译器，这个是一个 Outlet 变量。

（2）weak 代表弱引用，若弱引用的对象被销毁后，弱引用的指针会被清空。

（3）var 代表建立一个变量。

（4）titleLabel 代表变量的名称。

（5）UILabel！代表变量的类型为 UILabel 类，其后的"！"代表进行强制解析。程序员需要保证这个变量不能为 nil。

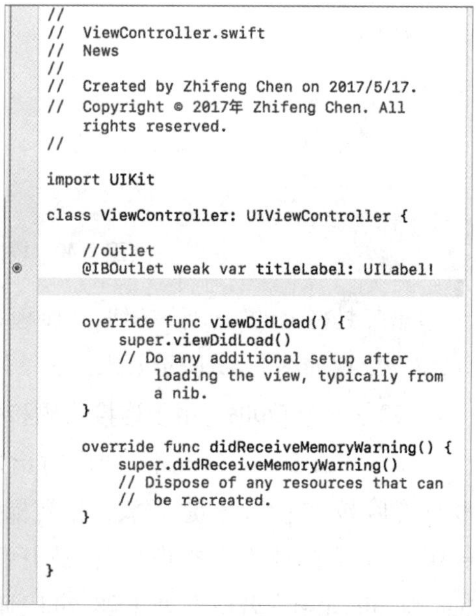

图 2-43　Outlet 的代码

10. 使用 Outlet，修改其对应 Label 的文字颜色

在 ViewController 类中，找到 viewDidLoad( ) 函数，在这个函数中输入以下代码，运行结果如图 2-44 所示。

```
titleLabel.text = "头条新闻"
titleLabel.textColor = UIColor.white
titleLabel.backgroundColor = UIColor.red
```

通过 Outlet 所连接的变量，可以修改其文字内容（text）、文字颜色（textColor），以及背景色（backgroundColor）等属性。

图 2-44　通过 Outlet 来修改文字内容和颜色

11. 拖放 UIImageView 到界面中，并调整大小

从对象库中拖放 UIImageView 到界面中，然后调整大小并设置 Autoresizing，如图 2-45 所示。

图 2-45　界面中的 UIImageView

12. 建立 UIImageView 的 Outlet

进入 Assistant Editor 状态，此时同时出现界面和代码。这样用户可以在界面区域选择 UIImageView，然后按住【Ctrl】键不放，用鼠标拖动，会出现一条蓝色橡皮筋，拖到 ViewController 类的代码中，具体可以在 viewDidLoad( ) 函数上方空白位置，放开鼠标左键，弹出如图 2-46 所示对话框，输入名称 name 为 titleImageView。

图 2-46　UIImageView 的 Outlet

（1）@ IBOutlet 告知编译器，这个是一个 Outlet 变量。

（2）weak 代表弱引用，若弱引用的对象被销毁后，弱引用的指针会被清空。

（3）var 代表建立一个变量。

（4）titleImageView 代表变量的名称。

（5）UIImageView！代表变量的类型为 UIImageView 类，其后的"！"代表进行强制解析。程序员需要保证这个变量不能为 nil。

13. 将图片资源拖入工程中

iOS 支持的标准图片格式为 png，当然 jpg 等常用文件格式也支持。为了能正常使用图片资源，需要在 Xcode 中把图片资源文件直接拖入到工程所在的文件夹，参数设置如图 2-47 所示。

14. 使用 Outlet，让 UIImageView 显示指定图片文件

在 ViewController 类中，找到 viewDidLoad( ) 函数，在这个函数中输入以下代码，运行结果如图 2-48 所示。

```
let image = UIImage(named:"kunshan.jpg")
titleImageView.image = image
```

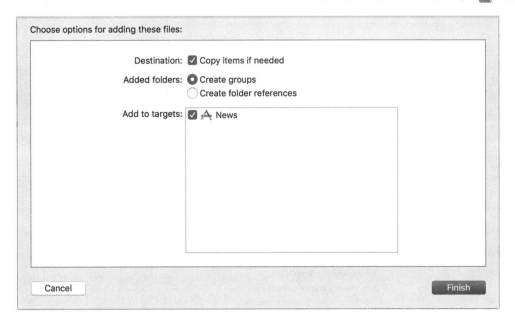

图 2-47　导入资源图片时的参数设置

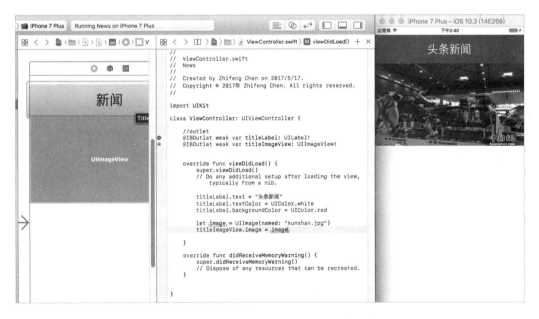

图 2-48　UIImageView 显示指定图片文件

在上述设计完成后,可以通过查看工程的连接情况,如 outlet,包括 action 的情况,如图 2-49所示。

2.3.2　Action

Actio 是一种特别的函数,在代码中对应的变量前面加上 IBAction 来修饰标记,就是告诉编译器,在界面事件发生后,需要一段代码来调用一个方法,响应这个操作。IBAction 就是用来标记代码中这个方法的。

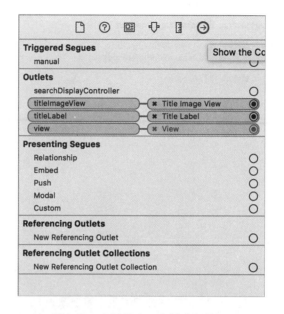

图 2-49 查看整个工程的连接情况

Action 要和 nib 文件中的一个界面对象关联起来,一旦建立了连接,就说当前这个标记了 IBAction 的方法是 nib 文件(也就是界面)中某对象所需要响应的事件。

下面以开发一个猜数字游戏运行(见图 2-50)为例:用户单击"游戏开始"按钮,系统生成一个一定范围内的随机整数,用户滑动滑杆,可以看到数值在不断变化,确定数值,然后单击"猜一猜"按钮,看这个数值和随机数是否相同,提示太大或者太小或者相等,每猜一次,就累加一次,看多少次以后用户猜对这个随机数。

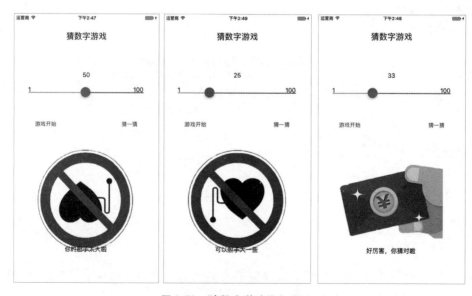

图 2-50 猜数字游戏运行界面

1. 项目界面规划

根据分析，本游戏需要 3 个 Action 事件：游戏开始、猜一猜、拖动滑杆(Slider)数值发生改变。

同样，本游戏需要 4 个 Outlet 变量：滑杆数值的显示(UILabel)、滑杆数值(UISlider)、结果图片显示(UIImageView)、结果提示信息(UILabel)。

猜数字游戏界面设计如图 2-51 所示。

图 2-51　猜数字游戏设计界面

2. 新建工程 Guess，保存到桌面

新建一个 Xcode 工程，名称为 Guess，保存到桌面上，和第一个 iOS 程序的步骤一致。

3. 打开 Main.storyboard，找到 View Controller

打开 Main.storyboard，选择 View Controller，单击文件查看器(File Inspector)，找到两个选项：Use Auto Layout 和 Use Trait Variations。

4. 取消 User Auto Layout 和 Use Trait Variations 功能

在取消这两个功能的时候，会出现一个提示框，选择同意即可。

5. 拖放组件到界面中，调整大小、位置和 Autoresizing 等

用户在对象库(Object Library)中找到 UILabel(5 个)、UISlider(1 个)、UIButton(2 个)、UIImageView(1 个)，将其拖放到界面的合适位置，进入尺寸查看器(Size Inspector)，调整的大小(width 和 height)，设置 Autoresizing 功能，以及 UISlide 数值的范围(1~100)和当前数值 50，如图 2-52 所示。

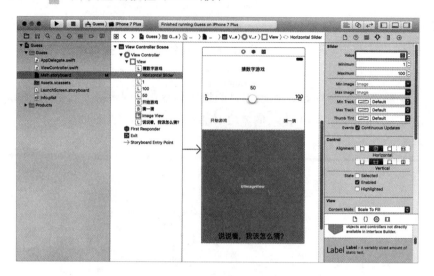

图 2-52　拖放组件，设置大小和 Autoresizing 等

6. 运行查看一下进展

单击工具栏上的"运行"按钮，并将模拟器进行向左旋转（选择 Hardware→Rotate Left 命令），将会出现如图 2-53 所示效果。

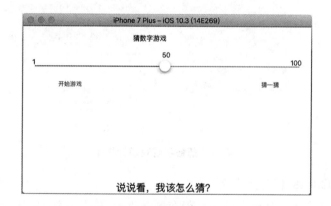

图 2-53　旋转后的运行界面

7. 建立一个 Action 和 outlet，实现滑杆拖动的时候显示数值

在 Assistant Editor 情况下，会同时出现界面和代码。用户可以在界面区域选择 Label，然后按住【Ctrl】键不放，用鼠标拖放到 ViewController 类的代码中，具体可以在 viewDidLoad() 函数上方空白位置，放开鼠标左键，弹出对话框，输入名称 Name 为 sliderValueLabel。

然后选择 UISlider 组件，按住【Ctrl】键不放，用鼠标拖放到 ViewController 类的代码中，具体可以在 viewDidLoad() 函数上方空白位置，放开鼠标左键，弹出对话框：Connection 选择 Action，Name 填写 onValueChanged：，Type 选择 UISlider，将会出现如图 2-54 所示效果。

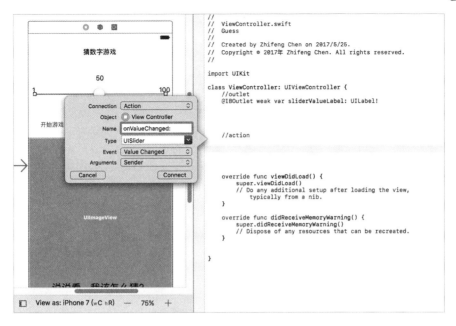

图 2-54　Action 的填写

8. 查看 Action 相关的代码和连接关系

如图 2-55 所示，可以获知这个 Action 代表的是一个函数，其函数名称为 onValueChanged(_ sender:UISlider)，函数有一个参数 sender，其类型为 UISlider。这个 Action 前面还有一个实心的圆圈，将鼠标指针放到这个圆圈上，相应界面的那个 Slider 会变颜色，表示连接已经存在。

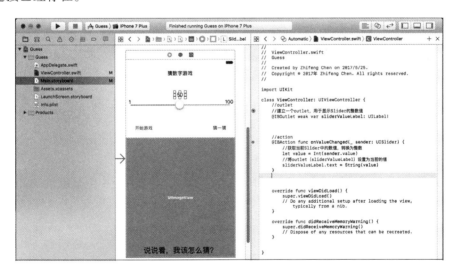

图 2-55　Action 的代码

（1）@IBAction func 告知编译器，这个是一个 Action 函数。

（2）func 是函数定义的标示。

(3) onValueChanged 代表函数名称。

(4) (_ sender:UISlider) 代表函数有一个参数,参数名称为 sender,其类型为 UISlider。

在这个 Action 事件中,需要从 Slider 获取当前的数值,并转换为整数,因为允许的范围是 1~100 的整数,然后将 outlet 变量 sliderValueLabel 中的内容设置为该数值,当然也需要采用 String() 实现将整数转换为字符串。

9. 建立其他 Action 和 Outlet,建立一个类的成员变量

在 Assistant Editor 情况下,继续通过选中组件,然后按住【Ctrl】键,用鼠标拖放到合适位置,如图 2-56 所示,依次完成 4 个 outlet 和 3 个 action 的设置,以及一个类的成员变量 keyCode,用于保存需要用户猜测的整数值。

```
import UIKit
class ViewController: UIViewController {
    //variable
    var keyCode : Int = 0

    //outlet
    //建立一个outlet, 用于显示Slider的整数值
    @IBOutlet weak var sliderValueLabel: UILabel!
    //建立一个outlet, 用于获取Slider中的数值
    @IBOutlet weak var valueSlider: UISlider!
    //建立一个outlet, 用于显示一个图片
    @IBOutlet weak var resultImageView: UIImageView!
    //建立一个outlet, 用于显示一个文本提示信息
    @IBOutlet weak var resultLabel: UILabel!

    //action 响应滑杆移动事件
    @IBAction func onValueChanged(_ sender: UISlider) {
        //获取当前Slider中的数值, 转换为整数
        let value = Int(sender.value)
        //将outlet (sliderValueLabel) 设置为当前的值
        sliderValueLabel.text = String(value)
    }

    //action 响应按钮触摸事件, 开始游戏
    @IBAction func onStartGame(_ sender: UIButton) {
    }

    //action 相应按钮触摸事件, 开始猜数字是否正确, 提示偏差
    @IBAction func onGuessed(_ sender: UIButton) {
    }
}
```

图 2-56 类成员变量,以及 outlet 和 action

10. 输入代码,实现相应的功能

根据系统设计功能:用户单击"游戏开始"按钮,系统生成一个一定范围内的随机整数,实现代码如下:

```
//action 响应按钮触摸事件,开始游戏
@ IBAction func onStartGame( _ sender:UIButton){
//生成一个1~100范围内的整数值
    keyCode = Int(arc4random()% 100 + 1)
}
```

根据系统设计功能:用户单击"猜一猜"按钮,看这个数值和系统保存的随机整数是否相同,提示太大或者太小或者相等。实现代码如下:

```
//action 相应按钮触摸事件,开始猜数字是否正确,提示偏差
@IBAction func onGuessed(_ sender:UIButton){
    let currentCode = Int(sliderValueLabel.text!)
    if keyCode > currentCode! {
        resultLabel.text = "你可以胆子大一些"
    }
    else if keyCode < currentCode! {
        resultLabel.text = "你胆子太大了一点"
    }
    else if keyCode == currentCode {
        resultLabel.text = "你太厉害啦!!!"
    }
}
```

11. 将图片导入到工程中,修改"猜一猜"的功能

为了能正常使用图片资源,需要在 Xcode 中,将图片资源文件直接拖入到工程所在的文件夹,然后修改"猜一猜"的代码,如图 2-57 所示。

注意:为了文件的整齐和归类,一般可以建立一个文件夹 pics,将所需要的图片文件都拖放到这个文件夹中,然后再将这个文件夹拖放到工程中。

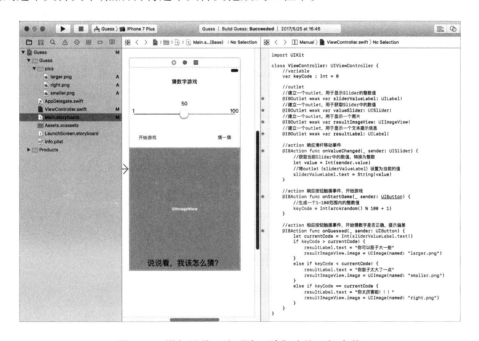

图 2-57 增加图片,让"猜一猜"功能更加完善

2.4 Playground 环境

Playground 是苹果在 Xcode 中添加的新功能。使用 Xcode 创建工程编写和运行程序,

目的是为了使最终的程序编译和发布,而使用 Playground 的目的是为了学习、测试算法、验证想法和可视化地看到运行结果。

用户可以直接在 Playground 中输入代码,快速熟悉各种代码的用法,调试各种函数,完成各种功能,然后就可以将代码直接使用到 Xcode 的工程中,大大节省开发调试时间。

2.4.1 Playground 的使用

通常情况下,直接在 Playground 上面写代码,然后编译器会实时编译代码,并将结果显示出来。

但是这也会产生一个问题,如果写了一个函数,或者自定义了一个 view,这部分代码一般情况下是不会变的,而编译器却会一次又一次地去编译这些代码,最终的结果就是导致效率降低。

这时,Sources 目录就派上用场了,使用"Cmd + 1"打开项目导航栏(Project Navigator),可以看到一个 Sources 目录。放到此目录下的源文件会被编译成模块(Module)并自动导入到 Playground 中,并且这个编译只会进行一次,而非每次输入一个字母时就编译一次,这将会大大提高代码执行的效率。

注意:由于此目录下的文件都是被编译成模块导入的,只有被设置成 public 的类型、属性或方法才能在 Playground 中使用。

由于 Playground 并没有使用沙盒机制,所以无法直接使用沙盒来存储资源文件。

但是,这并不意味着在 Playground 中没办法使用资源文件,Playground 提供了两个地方来存储资源:一个是每个 Playground 都独立的资源;另一个是所有 Playground 都共享的资源。

在打开的项目导航栏中可以看到有一个 Resources 目录,放置到此目录下的资源是每个 Playground 独立的。

这个目录的资源是直接放到 mainBundle 中的,可以使用如下代码来获取资源路径:

```
if let path = Bundle.main.path(forResource:"example", ofType:"json"){
// do something with json
//...
}
```

如果是图片文件,也可以直接使用 UIImage(named:)来获取。共享资源的目录是放在用户的 Documents 目录下的。在代码中可以直接使用 XCPlaygroundSharedDataDirectoryURL 来获取共享资源目录的 URL(需要先导入 XCPlayground 模块)。

```
import XCPlaygroud let sharedFileURL = XCPlaygroundSharedDataDirectoryURL.URLByAppendingPathComponent("example.json")
```

注意:需要创建 ~/Documents/Shared Playground Data 目录,并将资源放到此目录下,才能在 Playground 中获取到。

编写第一个应用程序 第 2 章

1. 启动 Xcode

在 macOS 中找到 Xcode，然后运行。启动 Xcode 的常见方法有：从 Dock 中找到 Xcode 图标；从 Launchpad 中找到 Xcode 图标；在 Spotlight 搜索中输入 Xcode，通过查找方式，找到 Xcode，双击然后即可运行，如图 2-58 所示。

图 2-58 从 Spotlight 搜索中运行 Xcode

2. 选择新建工程（Create a new Xcode project）

Xcode 的欢迎界面有 3 个选项：新建一个 playground 程序、新建一个 Xcode 工程和检查一个存在的工程。此处选择新建一个 playground，即第 1 项，如图 2-59 所示。从欢迎界面，还可以了解到当前 Xcode 的版本为 8.3.2。

图 2-59 新建工程（Get started with a playground）

3. 设置 Playground 的参数（Choose options）

Playground 参数设置比较简单：名称（Name），默认为 MyPlayground；开发平台（Platform），选择 iOS，如图 2-60 所示。

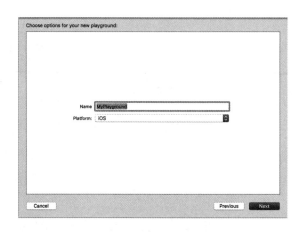

图 2-60　设置 Playground 的参数

4. 选择存放的目录路径

一般从左边菜单选择保存的位置在桌面，然后单击生成（Create）按钮，如图 2-61 所示。

图 2-61　选择项目保存的目录路径

5. Playground 运行界面

Playground 程序运行如图 2-62 所示，其中①区域是代码编写视图；②区域是控制台视图，使用 print 等日志函数将结果输出到控制台，可以通过左下角的按钮隐藏和显示控制台；③区域是时间轴视图；④区域是文件目录结构（图片等文件就是拖放到这个位置）。

编写第一个应用程序 第 2 章

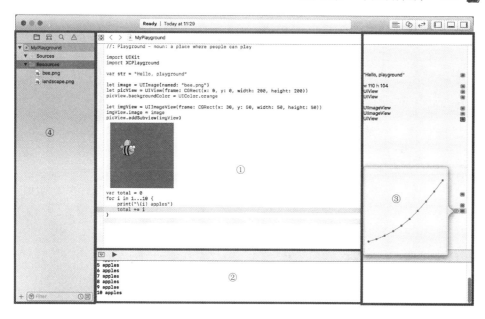

图 2-62　Playground 运行界面

2.4.2　Swift Playground

Swift Playground 是由苹果公司在 2016 年 6 月 14 日苹果全球开发者大会上发布的 iPad 应用,它通过一种老少皆宜的编程语言,让用户不用是程序员也能参与到编程当中。

在 iPad 上,Swift Playground 应用通过各种游戏来让小孩子享受编程带来的乐趣,使编程成为人们今后生活的必备技能,如图 2-63 所示。

图 2-63　iPad 上的 Swift Playground 应用

Swift Playground 的设计比较简单,主要分为 8 个大类的功能:let 语句、var 语句、for 循环、while 循环、repeat 循环、if 条件、switch 条件和自定义 func 等,如图 2-64 所示。

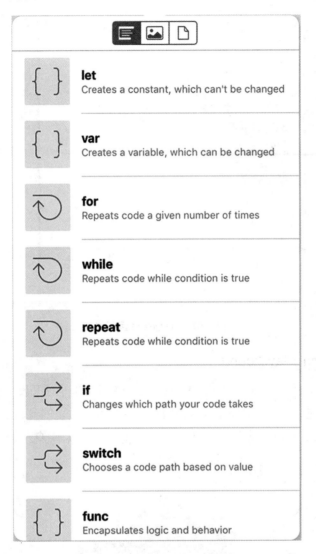

图 2-64　Swift Playground 的功能

第 3 章

听声音识动物

通过本章将学会根据声音的不同来认识动物。涉及的对象有太阳、小马、母牛、孔雀、蜜蜂、小鸟等。

3.1 功能简介

主要功能:太阳随机地变化大小,各种动物在触摸的时候会发出对应的声音,还有一个小兔子在蹦蹦跳跳,如图 3-1 所示。整个工程包括的组件有:太阳(UIImageView)、蹦蹦跳跳的小兔子(UIImageView)、蜜蜂(UIButton)、孔雀(UIButton)、小马(UIButton)、母牛(UIButton)以及背景图(UI Image View)。

因此,主要学习解决以下问题:图形按钮如何设置？声音与按钮上的动物如何对应播放？动画是怎么实现的？

首先应该把这个项目所需要的图片文件和音频资源文件,以及动画图片都准备好,如图 3-2 所示。

在 asset 文件夹中还包括两个子文件夹:animation 和 sound,其中 animation 文件夹中是一系列动画图片,如图 3-3 所示。

其中,sound 文件夹中是一系列声音文件,如图 3-4 所示。

图 3-1　听声音识动物工程

图 3-2　图片、声音和动画的所在文件夹 asset

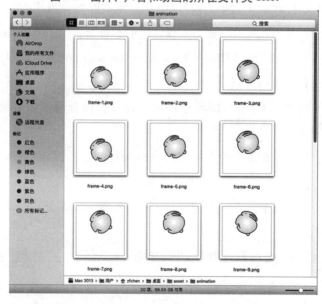

图 3-3　文件夹 animation 中的系列图片

听声音识动物　第 3 章

图 3-4　声音文件

3.2　音频播放

在 iOS 中音频播放从形式上可以分为音效播放和音乐播放。前者主要指的是一些短音频播放，通常作为点缀音频，对于这类音频不需要进行进度、循环等控制。后者指的是一些较长的音频，通常是主音频，对于这些音频的播放通常需要进行精确的控制。在 iOS 中播放两类音频原来分别使用 AudioToolbox 和 AVFoundation 来完成音效和音乐播放，现在都整合到 AVFoundation 中了。

音效，又称"短音频"，通常在程序中的播放时长为 1～2 s，在 APP 开发的过程中添加音效，往往能起到点缀效果，提升整体用户体验。下面简单介绍下 swift 中音效的播放以及对系统方法的封装。播放音效相关的 API 封装在 AVFoundation 框架中，一般来说只需要简单的三部，就能实现音效的播放。播放音效的步骤如下：

（1）定义一个 SystemSoundID。

（2）根据某一个音效文件，给 soundID 进行赋值。

（3）播放音效。

如果播放较大的音频或者要对音频有精确的控制，则 System Sound Service 可能就很难满足实际需求，通常这种情况会选择使用 AVAudioPlayer 来实现。AVAudioPlayer 可以看成一个播放器，它支持多种音频格式，而且能够进行进度、音量、播放速度等控制。AVAudio-Player 的使用比较简单：

（1）初始化 AVAudioPlayer 对象，此时通常指定本地文件路径。

（2）设置播放器属性，例如重复次数、音量大小等。

（3）调用 play 方法播放。

3.2.1　声音播放方法（林中鸟鸣）

在实现"听声音识动物"这个功能前，先做一个小项目"林中鸟鸣"，主要以实现采用一个按钮触摸事件来播放一个音频文件。

1. 界面规划

根据"林中鸟鸣"项目的实际运行情况,设计界面如图 3-5 所示。

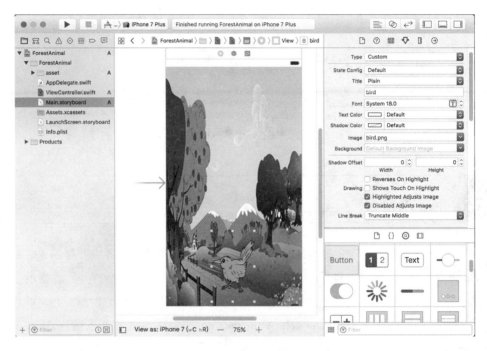

图 3-5　林中小鸟的鸣叫界面

通过分析可以看到,界面上的各个元素可以分解为两大类:UIButton 和 UIImageView,其中背景为一个 UIImageView,小鸟为一个 UIButton(注意标题为 bird),具体如图 3-6 所示。

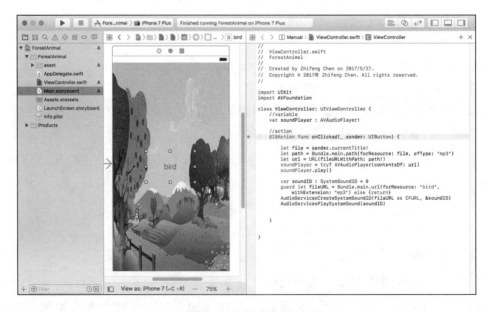

图 3-6　界面元素分解

2. 新建工程 ForestAnimal,保存到桌面

新建一个 Xcode 工程,名称为 ForestAnimal,保存到桌面上,和其他 iOS 程序的步骤一致。

3. 打开 Main. storyboard,找到 View Controller

打开 Main. storyboard,选择 View Controller,点击文件查看器(File Inspector),找到两个选项:Use Auto Layout 和 Use Trait Variations。

4. 取消 User Auto Layout 和 Use Trait Variations 功能

在取消这两个功能时,会出现一个提示,选择同意即可。

5. 把准备好的资源文件夹拖放到工程中

在拖放资源文件夹时,会出现一个提示,如图 3-7 所示选择即可。注意,千万不要选择 Create folder references 这个选项。

6. 拖放 UIImageView 到界面中,设置大小、图片和 Autoresizing

图 3-7 资源文件夹拖放进工程的选项设置

用户在对象库(Object Library)中找到 UIImageView,将其拖放到界面的合适位置,进入尺寸查看器(Size Inspector),调整 UIImageView 的大小(width 和 height),占满整个屏幕,将 image 设置为 background. png 文件,即背景,最后设置 Autoresizing 功能,如图 3-8 所示。

图 3-8 设置 UIImageView 的大小和 Autoresizing 功能

7. 运行查看一下进展

单击工具栏上的"运行"按钮,并将模拟器进行向左旋转(选择 Hardware→Rotate Left 命令),将会出现如图 3-9 所示效果。

图 3-9 模拟器旋转后的效果

8. 拖放一个 UIButton,选择图片,设置标题和 Autoresizing

用户拖放一个 UIButton 到界面的合适位置,如图 3-10 所示。在右侧找到 Image,可以直接在里面设置按钮的图片,现在选择小鸟图片 bird,同时修改其 Title 为 bird,最后不要忘记设置 Autoresizing。

图 3-10 拖放 UIButton 到界面中

9. 建立 UIButton 的 Action,输入相关代码

选择按钮,然后按住【Ctrl】键不放,拖放到代码空白处,在弹出的对话框中,选择 Action,输入名称 onClicked,选择 UIButton,确定后生成对应的函数,如图 3-11 所示。

图 3-11 小鸟按钮与声音播放代码

UIButton 的 Action 中就是播放声音的代码,具体如下:

```
//音乐文件播放
import UIKit
//1. 导入声音支持库
import AVFoundation
//添加到 self.view 上才会显示出来
class ViewController:UIViewController {
    //2.variable 建立一个类的成员变量,保存声音播放器变量
    var soundPlayer:AVAudioPlayer!
    //action 响应按钮事件
    @IBAction func onClicked(_ sender:UIButton){
        //获得当前按钮的 title,赋值给常量 file
        let file = sender.currentTitle!
        //将声音文件转化为 Bundle 中的路径
        let path = Bundle.main.path(forResource:file, ofType:"mp3")
        //将路径转化为 URL
        let url = URL(fileURLWithPath:path!)
        //根据 URL 建立 AVAudioPlayer 的实例,赋值给 soundPlayer 变量
        soundPlayer = try? AVAudioPlayer(contentsOf:url)
        //在实例变量中,调用 play() 函数
        soundPlayer.play()
    }
}
```

10. 运行欣赏声音的播放效果

运行工程，触摸按钮，即可播放声音，如图 3-12 所示。

3.2.2 声音文件制作

现在可以采用 macOS 自带的系统软件 Quick Time Player 来录音，这样可以制作拥有自己声音的软件。

1. 在 Launchpad 中找到并运行 Quick Time Player

打开 Launchpad，找到其他 Other 文件夹，打开 Other，可以看到 QuickTime Player，如图 3-13 所示。

图 3-12 林中鸟鸣的运行效果

图 3-13 查找 QuickTime Player 图标

2. 在"文件"菜单中找到"新建音频录制"

运行 QuickTime Player，在系统菜单中找到"文件"，单击后可以看到"新建音频录制"命令，如图 3-14 所示。

图 3-14 "文件"菜单中的"新建音频录制"命令

3. 打开"音频录制"窗口

选择"新建音频录制"命令,打开"音频录制"窗口,如图3-15所示。

4. 单击"录制"按钮,开始录制音频

单击红色"录制"按钮,开始录制,显示当前时长和文件大小,如图3-16所示。

图 3-15　音频录制窗口　　　　　　　　　图 3-16　开始录制音频,显示相关信息

5. 录制完成,进行播放

音频录制完成后,可以单击"播放"按钮听一下音频,如图3-17所示。

6. 音频文件保存菜单

选择Quick Time Player的菜单"文件"→"存储"命令,如图3-18所示。

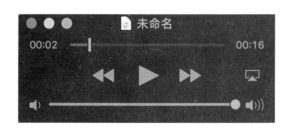 

图 3-17　录制完成,播放音频　　　　　　图 3-18　选择"文件"→"存储"命令

7. 设置保存的文件名称和位置

设置保存的文件名和位置,如图3-19所示。

图 3-19　设置保存的文件名和位置

3.3 动画播放

动画是将静止的画面变为动态的艺术，实现由静止到动态，主要是靠人眼的视觉残留效应。利用人的这种视觉生理特性可制作出具有高度想象力和表现力的动画影片。通过不断变化的图片，通过视觉暂留，形成连贯的动作，如图 3-20 所示。

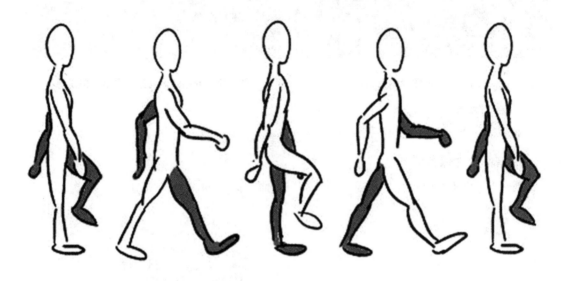

图 3-20　动画的一帧一帧图片

动画的英文有很多表述，如 Animation、Cartoon、Animated Cartoon、Cameracature。其中较正式、较常用的 Animation 一词源自于拉丁文字根 Anima，意思为"灵魂"，动词 Animate 是"赋予生命"的意思，引申为使某物活起来的意思。所以，动画可以定义为使用绘画的手法，创造生命运动的艺术。

从表面上看，电影胶片把一堆画面串在一条塑料胶片上。每一个画面称为一帧，代表电影中的一个时间片段。这些帧的内容总比前一帧有稍微的变化，这样，当电影胶片在投影机上放映时就产生了运动的错觉：每一帧都很短并且很快被另一个帧所代替，这样就产生了运动。

iOS 动画没什么不同，就像一个运动的画片一样，它包括许多独立的帧，每一帧都与前一帧略有不同。当移动时间轴上的播放头或放映电影时，用户在场景上所看到的就是每帧的图形内容，每帧图形的内容略有变化，当帧以足够快的速度放映时就会产生运动的错觉。

其原理都是制作一张张静态的图片，然后迅速地播放出来，利用人眼的视觉暂留，形成动画效果。

在 iOS 开发中,常见的动画主要有:UIImageView 所支持的系列图片动画;UIView 支持的旋转、平移、缩放和透明度变化等动画。

3.3.1 UIImageView 动画播放方法

1. UIImageView 系列图片动画方法一

如果现在工程中拥有一系列的动画图片其文件名有规律(例如,分别为 frame-1.png, frame-2.png,frame-3.png,…,frame-19.png),则可以采用以下程序代码来播放这个系列图片动画:

```
let iv = self.view.viewWithTag(4)! as! UIImageView
//UIImage 的一个方法 animatedImageNamed()可以调入一系列图片
//每个图片出现的间隔时间为 0.5s
let img = UIImage.animatedImageNamed("frame-",duration:0.5)
//设置动画
iv.image = img
```

2. UIImageView 系列图片动画方法二

如果现在工程中拥有一系列的动画图片,其文件名有规律(例如,分别为 frame-1.png, frame-2.png,frame-3.png,…,frame-19.png),则可以将这些文件添加到数组中,采用以下程序代码来播放这个系列图片动画:

```
var imgs:Array<UIImage> = []
for i in 1...19 {
    let img = UIImage(named:"frame-\(i).png")!
    imgs.append(img)
}
let imgView = self.view.viewWithTag(4)as! UIImageView
imgView.animationImages = imgs
imgView.animationDuration = 0.8
imgView.startAnimating()
```

3.3.2 UIView 动画播放方法

1. UIView 的旋转动画

如果现在工程中有一个组件,其 tag 为 4,则可以采用以下程序代码来旋转这个组件,本例中旋转 360°:

```
let iv = self.view.viewWithTag(4)!
UIView.animate(withDuration:2, animations:{
    iv.transform = iv.transform.rotated(by:CGFloat(360))
})
```

2. UIView 的平移动画

如果现在工程中有一个组件,其 tag 为 4,则可以采用以下程序代码来平移这个组件:

```
let iv = self.view.viewWithTag(4)!
UIView.animate(withDuration:2, animations:{
        iv.frame.origin.x + =100
if iv.frame.origin.x > self.view.frame.size.width {
        iv.frame.origin.x = 0
    }
})
```

3. UIView 的缩放动画

如果现在工程中有一个组件,其 tag 为 4,则可以采用以下程序代码来缩放这个组件,本例中缩小为原来的 0.8:

```
let iv = self.view.viewWithTag(4)!
UIView.animate(withDuration:2, animations:{
      iv.transform = CGAffineTransform(scaleX:0.8, y:0.8)
})
```

4. UIView 的透明度动画

如果现在工程中有一个组件,其 tag 为 4,则可以采用以下程序代码将这个组件的透明度设置为 0.1:

```
let iv = self.view.viewWithTag(4)!
UIView.animate(withDuration:2, animations:{
     iv.alpha = 0.1
})
```

3.4 功能实现

在前面"林中鸟鸣"的基础上,进一步实现整个"听声音识动物"功能。

3.4.1 在"林中鸟鸣"基础上实现全部功能

1. 复制 UIButton

选中现有的 UIButton(小鸟),按住【Alt】键不放,拖放鼠标,可以看到复制了一个一模一样的按钮,如图 3-21 所示。

2. 修改 UIButton 属性

主要修改的内容有两项:Title 修改为 bee,Image 修改为 bee.png,如图 3-22 所示。此时,看到界面上增加了一只可爱的小蜜蜂。

3. 运行并查看小蜜蜂的声音

运行后,可以触摸小鸟和小蜜蜂,会发出不同的声音。这是为什么?此时并没有为小蜜蜂建立相应的 action,但是将两个 UIButton 都连接到了一个 Action,如图 3-23 所示。

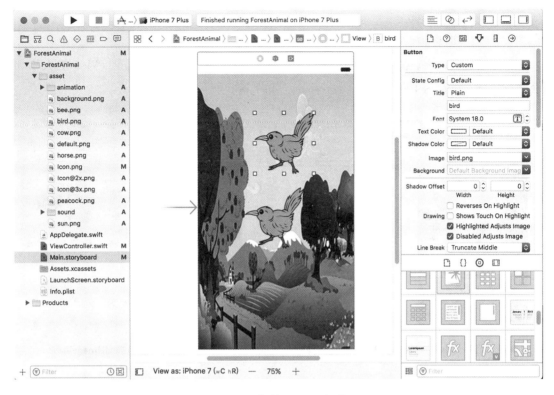

图 3-21　复制 UIButton 组件

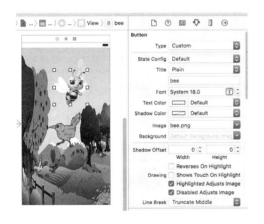

图 3-22　修改 UIButton 的属性

图 3-23　两个 UIButton 都连接到了同一个 Action

4. 依次复制奶牛、小马和孔雀并修改属性

依次选中 UIButton，按住【Alt】不放，拖放到其他位置，这样多次重复，然后逐个选中这些 UIButton，分别设置属性：奶牛（cow，cow.png）、小马（horse，horse.png）和孔雀（peacock，peacock.png），如图 3-24 所示。

5. 运行并查看奶牛、小马和孔雀的声音

运行后，可以触摸小鸟、小蜜蜂、奶牛、小马和孔雀，会发出不同的声音。这样就基本完成了这个项目的声音部分。

6. 拖放 UIImageView，设置大小和 Autoresizing

根据动画的需要，拖放一个 UIImageView 到界面设计中，设置该 UIImageView 的大小和合适位置，tag 设置为 1001，包括 Autoresizing，如图 3-25 所示。

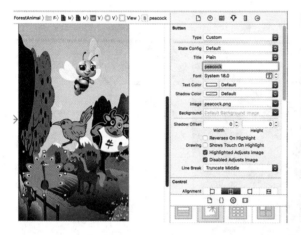
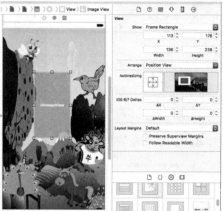

图 3-24　奶牛、小马、孔雀等 UIButton 复制并修改属性　　图 3-25　设置动画的 UIImageView

7. 动画代码的输入

如果希望在程序一开始运行时，就能自动实现动画，相应的程序代码必须放置到 view-DidLoad() 函数中，因为这个函数会在 View 显示的时候被自动调用。程序代码如图 3-26 所示。

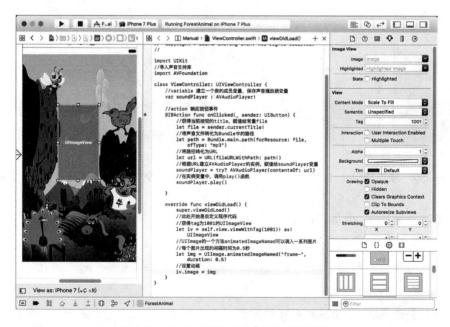

图 3-26　在 viewDidLoad() 中输入动画代码

8. 运行并查看动画效果

运行后可以看到一只小兔子在快乐地蹦蹦跳跳。

9. 应用 Timer 定时器实现 UIView 动画

时间控制器 Timer 可以实现定时器功能,即每隔一定时间执行具体函数,可以重复,也可以只执行一次。例如:

```
Timer.scheduledTimer(timeInterval:2.0, target:self, selector:#selector(doTimer), userInfo:nil, repeats:true)
```

其中,Timer 是系统对象,timeInterval:2.0 代表定时器 2 s 后触发,target:self 代表目标的归属,selector:#selector(doTimer)代表定时事件发生时调用函数 doTimer( ),userInfo:nil 代表用户信息为 nil,repeats:true 代表这个定时器要一直运行下去,如为 false,则只执行一次。

实现此功能的完整代码如下:

```
//  ViewController.swift
//  ForestAnimal
//  Created by Zhifeng Chen on 2017/5/27.
//  Copyright 2017 年 Zhifeng Chen. All rights reserved.

import UIKit
//导入声音支持库
import AVFoundation
class ViewController:UIViewController {
    //variable 建立一个类的成员变量,保存声音播放器变量
    var soundPlayer:AVAudioPlayer!
    //action 响应按钮事件
    @IBAction func onClicked(_ sender:UIButton){
        //获得当前按钮的 title,赋值给常量 file
        let file = sender.currentTitle!
        //将声音文件转化为 Bundle 中的路径
        let path = Bundle.main.path(forResource:file, ofType:"mp3")
        //将路径转化为 URL
        let url = URL(fileURLWithPath:path!)
        //根据 URL 建立 AVAudioPlayer 的实例,赋值给 soundPlayer 变量
        soundPlayer = try? AVAudioPlayer(contentsOf:url)
        //在实例变量中,调用 play()函数
        soundPlayer.play()
    }
    override func viewDidLoad(){
        super.viewDidLoad()
        //此处开始是自定义程序代码
        //获得 tag 为 1001 的 UIImageView
        let iv = self.view.viewWithTag(1001)! as! UIImageView
        //UIImage 的一个方法 animatedImageNamed 可以调入一系列图片
        //每个图片出现的间隔时间为 0.5s
        let img = UIImage.animatedImageNamed("frame-", duration:0.5)
        //设置动画
        iv.image = img
```

```
        //开启定时器,自动调用相关函数
        Timer.scheduledTimer(timeInterval:2.0, target:self, selector:#selector(doTimer),
userInfo:nil, repeats:true)
    }
    func doTimer(){
        let iv = self.view.viewWithTag(1001)!
        UIView.animate(withDuration:2, animations:{
            iv.transform = iv.transform.rotated(by:CGFloat(360))
        })
    }
}
```

3.4.2 问题与提高

在运行中发现,按钮上的标题文字在某些情况下会显示出来,特别是将屏幕旋转后,比较影响程序的效果。有没有更好的办法呢? 这就要通过组件的 tag 值来区分哪个组件发生了触摸事件,然后播放相应的音频文件,这样就不再需要在每个按钮上设置标题。

依次把蜜蜂、小鸟、奶牛、小马和孔雀的 tag 值设置为 1、2、3、4、5,把每个 UIButton 的 title 中的文字删除。程序进行相应的修改。

```
// ViewController.swift
// ForestAnimal
// Created by Zhifeng Chen on 2017/5/27.
// Copyright 2017 年 Zhifeng Chen. All rights reserved.

import UIKit
//导入声音支持库
import AVFoundation
class ViewController:UIViewController {
    //variable 建立一个类的成员变量,保存声音播放器变量
    var soundPlayer:AVAudioPlayer!
    //action 响应按钮事件
    @IBAction func onClicked(_ sender:UIButton){
        //获得当前按钮的 title,赋值给常量 file
        let index = sender.tag
        var file = ""
        switch index {
        case 1:
            file = "bee"
        case 2:
            file = "bird"
        case 3:
            file = "cow"
        case 4:
            file = "horse"
        case 5:
            file = "peacock"
        default :
            file = "bee"
        }
```

```
            //将声音文件转化为 Bundle 中的路径
            let path = Bundle.main.path(forResource:file, ofType:"mp3")
            //将路径转化为 URL
            let url = URL(fileURLWithPath:path!)
            //根据 URL 建立 AVAudioPlayer 的实例,赋值给 soundPlayer 变量
            soundPlayer = try? AVAudioPlayer(contentsOf:url)
            //在实例变量中,调用 play()函数
            soundPlayer.play()
        }
        override func viewDidLoad(){
            super.viewDidLoad()
            //此处开始是自定义程序代码
            //获得 tag 为 1001 的 UIImageView
            let iv = self.view.viewWithTag(1001)! as! UIImageView
            //UIImage 的一个方法 animatedImageNamed 可以调入一系列图片
            //每个图片出现的间隔时间为 0.5s
            let img = UIImage.animatedImageNamed("frame-", duration:0.5)
            //设置动画
            iv.image = img
            //开启定时器,自动调用相关函数
            Timer.scheduledTimer(timeInterval:2.0, target:self, selector:#selector(doTimer),
userInfo:nil, repeats:true)
        }
        func doTimer(){
            let iv = self.view.viewWithTag(1001)!
            UIView.animate(withDuration:2, animations:{
                iv.transform = iv.transform.rotated(by:CGFloat(360))
            })
        }
    }
```

3.5 拓展学习:纯代码编程

前面通过使用苹果公司推荐的 Storyboard 来构建项目,那么能否只通过代码,不进行可视化的组件 UI 拖放,来实现所有功能呢?下面就介绍一下如何通过纯代码来完成项目。

3.5.1 做好准备工作

和 Storyboard 一样,需要准备好所需要的各种资源文件,如声音、图片、动画等内容,并规划好各个组件的位置,如图 3-1~图 3-4 所示。

1. 新建工程 ForestCode,保存到桌面

新建一个 Xcode 工程,名称为 ForestCode,保存到桌面上,和其他 iOS 程序的步骤一致。

2. 将资源文件夹拖放到工程中

用户准备好自己的资源文件夹,然后在 Xcode 直接拖放到工程中,如图 3-27 所示。

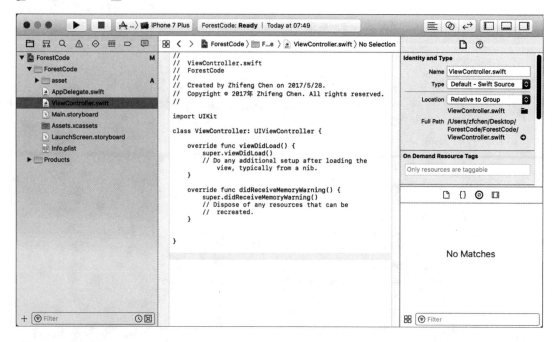

图 3-27　纯代码编程的准备工作

3.5.2　编写代码

在 iOS 中,其内部框架的各个类之间是存在一定调用顺序的,在某一个 ViewController 中,viewDidLoad( )是在界面初始化成功的时候被自动调用,因此,一般把需要初始化的界面代码书写在这个函数中是比较合适的。

1. 游戏背景的设置

本工程中,背景是通过 UIImageView 组件,设置全屏幕大小来实现的,这个组件所使用的背景图片为 background.png。

打开 ViewController.swift 这个文件,找到 viewDidLoad( )函数,将该函数的内容修改如下:

```
override func viewDidLoad(){
    super.viewDidLoad()
    // Do any additional setup after loading the view, typically from a nib.
    //获得当前屏幕的宽度
    let Width = self.view.frame.size.width
    //获得当前屏幕的高度
    let Height = self.view.frame.size.height
    //将图片文件 background.png 调入内存
    let img = UIImage(named:"background.png")
    //设置一个矩形框
    let rect = CGRect(x:0, y:0, width:Width, height:Height)
    //根据矩形框 rect 的大小,建立一个 UIImageView 对象
    let imgView = UIImageView(frame:rect)
```

```
        //设置UIImageView对象中的图片
        imgView.image = img
        //显示UIImageView
        self.view.addSubview(imgView)
        
}
```

这里面出现了 self.view 类型的每个实例都有一个隐藏的属性 self，等同于实例本身，但只要在一个方法中使用一个已知的属性或方法，即使没有明确地写出 self，Swift 也会假定是当前的，但是如果实例方法的某个参数名称与实例的某个属性名称相同，这种情况下参数名称享有优先权，这时需要 self 来区分参数名称和属性名称。

2. 蜜蜂按钮的添加

图形按钮是通过 UIButton 组件在按钮上设置图片来实现的。在本工程中，图形按钮有 5 个，所使用的背景图片分别为 bee.png、bird.png、cow.png、horse.png 和 peacock.png。

下面完成了一个蜜蜂按钮的添加，整个源代码如下：

```
import UIKit
class ViewController:UIViewController {
    override func viewDidLoad(){
        super.viewDidLoad()
        // Do any additional setup after loading the view, typically from a nib.
        //获得当前屏幕的宽度
        let Width = self.view.frame.size.width
        //获得当前屏幕的高度
        let Height = self.view.frame.size.height
        //将图片文件 background.png 调入内存
        let img = UIImage(named:"background.png")
        //设置一个矩形框
        let rect = CGRect(x:0, y:0, width:Width, height:Height)
        //根据矩形框 rect 的大小，建立一个 UIImageView 对象
        let imgView = UIImageView(frame:rect)
        //设置UIImageView对象中的图片
        imgView.image = img
        //显示UIImageView
        self.view.addSubview(imgView)
        //设置蜜蜂按钮所需要的矩形框
        let beeRect = CGRect(x:130, y:60, width:80, height:80)
        //建立蜜蜂按钮对象
        let beeBtn = UIButton(frame:beeRect)
        //设置蜜蜂按钮的 tag 值为 1
        beeBtn.tag = 1
        //设置蜜蜂按钮的图片为 bee.png
        beeBtn.setImage(UIImage(named:"bee.png"), for:UIControlState.normal)
        //设置蜜蜂按钮触摸事件所对应的函数为 onClicked(_:)
        beeBtn.addTarget(self, action:#selector(onClicked(_:)), for:UIControlEvents.touchUpInside)
        //显示蜜蜂按钮
```

```swift
        self.view.addSubview(beeBtn)
    }
    //按钮事件所对应的函数
    func onClicked(_ sender:UIButton){
        let tag = sender.tag
        print(tag)
    }
    override func didReceiveMemoryWarning(){
        super.didReceiveMemoryWarning()
        // Dispose of any resources that can be recreated.
    }
}
```

3. 其他按钮的添加

按照类似代码,完成其他图形按钮的添加,源代码如下:

```swift
override func viewDidLoad(){
    super.viewDidLoad()
    // Do any additional setup after loading the view, typically from a nib.
    //获得当前屏幕的宽度
    let Width = self.view.frame.size.width
    //获得当前屏幕的高度
    let Height = self.view.frame.size.height
    
    //将图片文件background.png调入内存
    let img = UIImage(named:"background.png")
    
    //设置一个矩形框
    let rect = CGRect(x:0, y:0, width:Width, height:Height)
    //根据矩形框rect的大小,建立一个UIImageView对象
    let imgView = UIImageView(frame:rect)
    //设置UIImageView对象中的图片
    imgView.image = img
    
    //显示UIImageView
    self.view.addSubview(imgView)
    
    //设置蜜蜂按钮所需要的矩形框
    let beeRect = CGRect(x:130, y:60, width:80, height:80)
    //建立蜜蜂按钮对象
    let beeBtn = UIButton(frame:beeRect)
    //设置蜜蜂按钮的tag值为1
    beeBtn.tag = 1
    //设置蜜蜂按钮的图片为bee.png
    beeBtn.setImage(UIImage(named:"bee.png"), for:UIControlState.normal)
    //设置蜜蜂按钮触摸事件所对应的函数为onClicked(_:)
    beeBtn.addTarget(self, action:#selector(onClicked(_:)), for:UIControlEvents.touchUpInside)
    //显示蜜蜂按钮
    self.view.addSubview(beeBtn)
```

```swift
        //设置小鸟按钮所需要的矩形框
        let birdRect = CGRect(x:280, y:180, width:80, height:80)
        //建立小鸟按钮对象
        let birdBtn = UIButton(frame:birdRect)
        //设置小鸟按钮的tag值为2
        birdBtn.tag = 2
        //设置小鸟按钮的图片为bird.png
        birdBtn.setImage(UIImage(named:"bird.png"), for:UIControlState.normal)
        //设置小鸟按钮触摸事件所对应的函数为onClicked(_:)
        birdBtn.addTarget(self, action:#selector(onClicked(_:)), for:UIControlEvents.touchUpInside)
        //显示小鸟按钮
        self.view.addSubview(birdBtn)
        //设置奶牛按钮所需要的矩形框
        let cowRect = CGRect(x:Width-100, y:Height-120, width:80, height:80)
        //建立奶牛按钮对象
        let cowBtn = UIButton(frame:cowRect)
        //设置奶牛按钮的tag值为3
        cowBtn.tag = 3
        //设置奶牛按钮的图片为cow.png
        cowBtn.setImage(UIImage(named:"cow.png"), for:UIControlState.normal)
        //设置奶牛按钮触摸事件所对应的函数为onClicked(_:)
        cowBtn.addTarget(self, action:#selector(onClicked(_:)), for:UIControlEvents.touchUpInside)
        //显示奶牛按钮
        self.view.addSubview(cowBtn)
        
        //设置小马按钮所需要的矩形框
        let horseRect = CGRect(x:Width-200, y:Height-100, width:80, height:80)
        //建立小马按钮对象
        let horseBtn = UIButton(frame:horseRect)
        //设置小马按钮的tag值为4
        horseBtn.tag = 4
        //设置小马按钮的图片为cow.png
        horseBtn.setImage(UIImage(named:"horse.png"), for:UIControlState.normal)
        //设置小马按钮触摸事件所对应的函数为onClicked(_:)
        horseBtn.addTarget(self, action:#selector(onClicked(_:)), for:UIControlEvents.touchUpInside)
        //显示小马按钮
        self.view.addSubview(horseBtn)
        
        //设置孔雀按钮所需要的矩形框
        let peacockRect = CGRect(x:90, y:Height-150, width:80, height:80)
        //建立孔雀按钮对象
        let peacockBtn = UIButton(frame:peacockRect)
        //设置孔雀按钮的tag值为5
        peacockBtn.tag = 5
        //设置孔雀按钮的图片为cow.png
        peacockBtn.setImage(UIImage(named:"peacock.png"), for:UIControlState.normal)
        //设置孔雀按钮触摸事件所对应的函数为onClicked(_:)
        peacockBtn.addTarget(self, action:#selector(onClicked(_:)), for:UIControlEvents.touchUpInside)
        //显示孔雀按钮
        self.view.addSubview(peacockBtn)
    }
```

程序运行效果如图 3-28 所示。

图 3-28　程序运行效果

4. 太阳图片的添加

根据设计界面，需要在初始化时添加一个 UIImageView，并将其图片设置为 sun.png，具体的代码应该添加在 viewDidLoad() 中。

```
//设置太阳所需要的矩形框
let sunRect = CGRect(x:Width-90, y:30, width:80, height:80)
//建立太阳图片的对象
let sunImageView = UIImageView(frame:sunRect)
//设置对象的 image 属性为 sun.png 图片
sunImageView.image = UIImage(named:"sun.png")
//显示太阳图片
self.view.addSubview(sunImageView)
```

5. 动画的添加

根据设计界面，需要在初始化时添加一个 UIImageView，实现动画效果，动画图片依次为 frame-1.png、frame-2.png、…、frame-19.png，具体的代码应该添加在 viewDidLoad() 中。

```
//设置动画所需要的矩形框
let aniRect = CGRect(x:190, y:230, width:100, height:180)
//建立一个空数组,其类型为 UIImage
var imgs:Array<UIImage> = []
//通过一个循环,把 frame-1.png 到 frame-19.png 这 19 个图片添加到数组中
for i in 1...19 {
    let each = UIImage(named:"frame-\(i).png")!
    imgs.append(each)
}
//建立 UIImageView 对象
let aniImageView = UIImageView(frame:aniRect)
```

```
//设置UIImageView的动画数组
aniImageView.animationImages = imgs
//开始动画
aniImageView.startAnimating()
//显示动画
self.view.addSubview(aniImageView)
```

6. 声音的添加

声音的支持首先需要导入库(import AVFoundation)，其次要建立一个类的成员变量 var soundPlayer:AVAudioPlayer!，最后需要将声音文件调入内存播放。

整个工程的完整代码如下：

```
//  ViewController.swift
//  ForestCode
//  Created by Zhifeng Chen on 2017/5/28.
//  Copyright 2017年 Zhifeng Chen. All rights reserved.

import UIKit
import AVFoundation
class ViewController:UIViewController {
    //建立一个类成员变量
    var soundPlayer:AVAudioPlayer!
    override func viewDidLoad(){
        super.viewDidLoad()
        // Do any additional setup after loading the view, typically from a nib.
        //获得当前屏幕的宽度
        let Width = self.view.frame.size.width
        //获得当前屏幕的高度
        let Height = self.view.frame.size.height
        //将图片文件background.png调入内存
        let img = UIImage(named:"background.png")
        //设置一个矩形框
        let rect = CGRect(x:0, y:0, width:Width, height:Height)
        //根据矩形框rect的大小，建立一个UIImageView对象
        let imgView = UIImageView(frame:rect)
        //设置UIImageView对象中的图片
        imgView.image = img
        //显示UIImageView
        self.view.addSubview(imgView)
        
        //设置蜜蜂按钮所需要的矩形框
        let beeRect = CGRect(x:130, y:60, width:80, height:80)
        //建立蜜蜂按钮对象
        let beeBtn = UIButton(frame:beeRect)
        //设置蜜蜂按钮的tag值为1
        beeBtn.tag = 1
        //设置蜜蜂按钮的图片为bee.png
        beeBtn.setImage(UIImage(named:"bee.png"), for:UIControlState.normal)
```

```swift
//设置蜜蜂按钮触摸事件所对应的函数为onClicked(_:)
beeBtn.addTarget(self, action:#selector(onClicked(_:)), for:UIControlEvents.touchUpInside)
//显示蜜蜂按钮
self.view.addSubview(beeBtn)
//设置小鸟按钮所需要的矩形框
let birdRect = CGRect(x:280, y:180, width:80, height:80)
//建立小鸟按钮对象
let birdBtn = UIButton(frame:birdRect)
//设置小鸟按钮的tag值为2
birdBtn.tag = 2
//设置小鸟按钮的图片为bird.png
birdBtn.setImage(UIImage(named:"bird.png"), for:UIControlState.normal)
//设置小鸟按钮触摸事件所对应的函数为onClicked(_:)
birdBtn.addTarget(self, action:#selector(onClicked(_:)), for:UIControlEvents.touchUpInside)
//显示小鸟按钮
self.view.addSubview(birdBtn)

//设置奶牛按钮所需要的矩形框
let cowRect = CGRect(x:Width-100, y:Height-120, width:80, height:80)
//建立奶牛按钮对象
let cowBtn = UIButton(frame:cowRect)
//设置奶牛按钮的tag值为3
cowBtn.tag = 3
//设置奶牛按钮的图片为cow.png
cowBtn.setImage(UIImage(named:"cow.png"), for:UIControlState.normal)
//设置奶牛按钮触摸事件所对应的函数为onClicked(_:)
cowBtn.addTarget(self, action:#selector(onClicked(_:)), for:UIControlEvents.touchUpInside)
//显示奶牛按钮
self.view.addSubview(cowBtn)

//设置小马按钮所需要的矩形框
let horseRect = CGRect(x:Width-200, y:Height-100, width:80, height:80)
//建立小马按钮对象
let horseBtn = UIButton(frame:horseRect)
//设置小马按钮的tag值为4
horseBtn.tag = 4
//设置小马按钮的图片为cow.png
horseBtn.setImage(UIImage(named:"horse.png"), for:UIControlState.normal)
//设置小马按钮触摸事件所对应的函数为onClicked(_:)
horseBtn.addTarget(self, action:#selector(onClicked(_:)), for:UIControlEvents.touchUpInside)
//显示小马按钮
self.view.addSubview(horseBtn)

//设置孔雀按钮所需要的矩形框
let peacockRect = CGRect(x:90, y:Height-150, width:80, height:80)
//建立孔雀按钮对象
let peacockBtn = UIButton(frame:peacockRect)
//设置孔雀按钮的tag值为5
peacockBtn.tag = 5
```

```swift
//设置孔雀按钮的图片为cow.png
peacockBtn.setImage(UIImage(named:"peacock.png"), for:UIControlState.normal)
//设置孔雀按钮触摸事件所对应的函数为onClicked(_:)
peacockBtn.addTarget(self, action:#selector(onClicked(_:)), for:UIControlEvents.touchUpInside)
//显示孔雀按钮
self.view.addSubview(peacockBtn)

//设置太阳所需要的矩形框
let sunRect = CGRect(x:Width-90, y:30, width:80, height:80)
//建立太阳图片的对象
let sunImageView = UIImageView(frame:sunRect)
//设置对象的image属性为sun.png图片
sunImageView.image = UIImage(named:"sun.png")
//显示太阳图片
self.view.addSubview(sunImageView)

//设置动画所需要的矩形框
let aniRect = CGRect(x:190, y:230, width:100, height:180)
//建立一个空数组,其类型为UIImage
var imgs:Array<UIImage> = []
//通过一个循环,把frame-1.png到frame-19.png这19个图片添加到数组中
for i in 1...19 {
    let each = UIImage(named:"frame-\(i).png")!
    imgs.append(each)
}
//建立UIImageView对象
let aniImageView = UIImageView(frame:aniRect)
//设置UIImageView的动画数组
aniImageView.animationImages = imgs
//开始动画
aniImageView.startAnimating()
//显示动画
self.view.addSubview(aniImageView)
}
//按钮事件所对应的函数
func onClicked(_ sender:UIButton){
    let tag = sender.tag
    print(tag)
    var file = ""
    switch tag {
    case 1:
        file = "bee"
    case 2:
        file = "bird"
    case 3:
        file = "cow"
    case 4:
        file = "horse"
```

```
        case 5:
            file = "peacock"
        default:
            file = ""
        }
        let path = Bundle.main.path(forResource:file, ofType:"mp3")
        let url = URL(fileURLWithPath:path!)
        soundPlayer = try? AVAudioPlayer(contentsOf:url)
        soundPlayer.play()
    }
    override func didReceiveMemoryWarning(){
        super.didReceiveMemoryWarning()
        // Dispose of any resources that can be recreated.
    }
}
```

3.5.3 屏幕旋转的处理

常见屏幕旋转的检测方法是利用手机内支持的通知监听消息实现的：

UIDeviceOrientationDidChangeNotification

当手机的物理方向发生变化时,将会发布这个通知,如果监听这个通知,那么在通知的相应方法里面就可以处理旋转过程中需要的操作。

```
//1. 在控制器的viewDidLoad()适当的地方注册通知监听者
//设置UIDevice方向变化监听的通知 NotificationCenter.default.addObserver(self, selector:#selector(receivedRotation(notification:)), name: NSNotification.Name.UIDeviceOrientationDidChange, object:nil)//

//2. 处理旋转过程中需要的操作
//UIDevice旋转方向通知监听触发的方法
func receivedRotation(notification:NSNotification){
    backgroundView.frame = CGRect(x:0, y:0, width:self.view.frame.size.width, height:self.view.frame.size.height)
}
```

第 4 章

找出"你我"不同

每个人都喜欢玩游戏,这样可以放松自己,那么能否开发一个自己的游戏,比如发挥自己的眼力,找出两种图片之间的区别呢?要求通过寻找图像之间的差异,看是否能在最快的时间内完成。

4.1 功能简介

场景:在悠闲中,在充满刺激恐怖的午夜里,在青青的草地上,以及蛙声一片的池塘边,放松心情。要求在场景中随机位置生成不同的小图片,这样每次在相同场景下,找不同的答案都是不一样的。这里需要准备 4 个 640×400 像素的场景图片(back01.png,…,back04.png)和 5 个 40×40 像素的小图片(bee.png、bird.png、flower.png、mogu.png 和 empty.png),其中有一个为透明的空白图片(empty.png),如图 4-1 所示。

主要功能:2 个按钮 UIButton(上一个和下一个)可以导航到不同的页面,每个页面包括上下 2 个 UIImageView,一个 UILabel 用于说明这个游戏的玩法(找出下面图片中的不同之处)。

这里采用了故事板 Storyboard 进行设计,一共包括 4 个页面,可以清楚地看到其中的各种关系,如图 4-2 所示。

图 4-1　4 个场景图片和五个小图片

图 4-2　找不同游戏的 4 个场景

4.2　故事板（Storyboard）

　　随着 iOS 的发展，可以说在 UI 制作上大家逐渐分化为了 3 种主要流派：使用代码手写 UI 及布局；使用单个 XIB 文件组织 viewController 或者 view；使用 StoryBoard 来通过单个或很少的几个文件构建全部 UI。

（1）手写代码：代码 UI 可以说具有最好的代码重用性。如果写一些可以高度重用的控件提供给其他开发者使用，那毫无疑问最好的选择应该是使用代码来完成UIView的子类。这样进一步修改和其他开发者在使用时，都会方便不少。

（2）XIB 模式：现在使用越来越少。其实 IB(Interface Builder)和 XIB 是从 iOS SDK 初次面世开始就是捆绑在开发者工具套装内的内容了。Xcode 4 之后被直接集成到了 Xcode 中成为了 IDE 的一部分。用 XIB 设计的一大目的其实是为了良好的 MVC(模型、视图、控制器)。一般来说，单个的 XIB 文件对应一个 ViewController，而对于一些自定义的 view，往往也会使用单个 XIB 并从 main bundle 进行加载的方式来载入，IB 帮助完成 view 的创建、布局和与 file owner 的关系映射等一些列工作。最大的问题在于 XIB 中的设置往往并非最终设置，在代码中将有机会覆盖在 XIB 文件中进行的 UI 设计。在不同的地方对同一个属性进行设置，之后将很难维护，因为其实 IB 还是有所局限的，它没有逻辑判断，也很难在运行时进行配置。

（3）StoryBoard：简单理解，可以把 StoryBoard 看作是一组 viewController 对应的 XIB，以及它们之间的转换方式的集合。在 StoryBoard 中不仅可以看到每个 ViewController 的布局样式，也可以明确地知道各个 ViewController 之间的转换关系。

现在 StoryBoard 面临的最大问题就是多人协作。因为所有的 UI 都定义在一个文件中，因此很多开发者个人或企业的技术负责人认为 StoryBoard 是无法进行协作开发的，其实这更多的是一种对 StoryBoard 的陌生所造成的误解。其实，整个项目可以使用多个StoryBoard 文件。当然，现在还有一些对于 StoryBoard 性能上的担忧，因为相对于单个 XIB 来说，Story-Board 文件往往更大，加载速度也相应变慢。

4.2.1　找不同游戏的制作

1. 建立一个工程 Differents

在 macOS 中找到 Xcode，然后运行。在 Xcode 的欢迎界面选择新建一个 Xcode 工程。从工程模版选择开发 iOS 应用程序，然后选择 Single View Application。在工程参数中主要包括输入工程的名称(Product Name)，填写 Differents，最后选择存放在桌面上。

2. 选择 Main. storyboard，拖放 UIImageView 组件等

从左边菜单单击 Main. storyboard 文件，在编辑区域会出现手机可视化设计界面。现在里面只有一个 View Controller 界面，从右边功能区域选择 2 个 UIImageView 组件、2 个 UIButton 和 1 个 UILabel，拖放到 View Controller 界面中，设置好每个组件的 Autoresizing，如图 4-3 所示。

为了方便地使用两个 UIImageView，可将上面的 UIImageView 的 tag 设置为 1001，下面的 tag 设置为 1002，这样可以通过函数 viewWithTag(1001)来找到下面的那个UIImageView，而不再需要通过 outlet 来处理，非常方便。

注意:为了每个组件的位置能自动排版到合适的位置,需要正确设置每个组件 Autoresizing 选项。

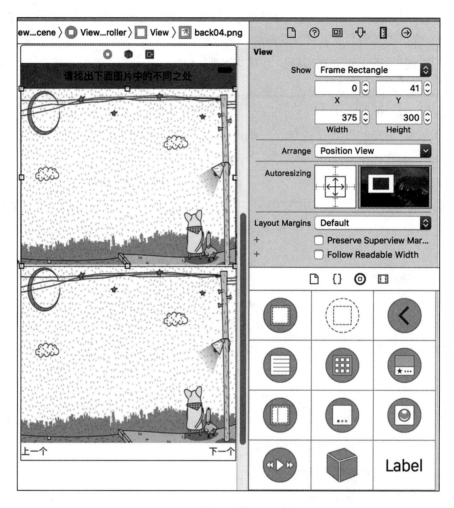

图 4-3 找不同的界面设计

3. 选择 ViewController.swift,输入代码

修改 ViewController.swift 中的代码,主要实现在上下场景图片添加 10 个小图片,起到混淆作用,然后再添加 3 个不同点,用户触摸不同点,则会输出提示信息。

程序代码如下:

```
//ViewController.swift
// Differents
// Created by ChenZhifeng on 17/6/4.
// Copyright © 2017 年 ChenZhifeng. All rights reserved.

import UIKit
class ViewController: UIViewController {
    //保存 4 个场景图片的 UIImage 数组
```

找出"你我"不同 第4章

```swift
var images : Array<UIImage> = [
    UIImage(named: "bee.png")!,
    UIImage(named: "bird.png")!,
    UIImage(named: "flower.png")!,
    UIImage(named: "mogu.png")!
]

//保存各个小图片的坐标数组,上下两个场景都具有这个小图片,起到混淆作用
var cords : Array<CGPoint> = []
//保存不同点所在坐标的数组
var errorCords : Array<CGPoint> = []

//保存当前场景的宽度和高度的变量Width、Height
var Width : CGFloat!
var Height : CGFloat!
override func viewDidLoad() {
    super.viewDidLoad()
    // Do any additional setup after loading the view, typically from a nib.
    //通过tag数值,找到上下两个场景图片的UIImageView
    let iv1 = self.view.viewWithTag(1001) as! UIImageView
    let iv2 = self.view.viewWithTag(1002) as! UIImageView
    //设置可以处理触摸事件
    iv1.isUserInteractionEnabled = true
    iv2.isUserInteractionEnabled = true
    //设置小图片可以出现的最大宽度和高度,为什么都要减去40?
    Width = iv1.frame.size.width - 40
    Height = iv1.frame.size.height - 40
    
    //生成十个随机数坐标(x,y),保存到cords数组中
    for _ in 0...9 {
        let x = CGFloat(arc4random() % UInt32(Width))
        let y = CGFloat(arc4random() % UInt32(Height))
        cords.append(CGPoint(x: x, y: y))
    }
    //生成3个产生不同点的位置坐标,保存到errorCords数组中
    for _ in 0...2 {
        let x = CGFloat(arc4random() % UInt32(Width))
        let y = CGFloat(arc4random() % UInt32(Height))
        errorCords.append(CGPoint(x: x, y: y))
    }
    //根据上面生成的10个坐标,在上下场景中,生成一样的效果,起到混淆作用
    for each in cords {
        let size = CGSize(width: 40, height: 40)
        let rect = CGRect(origin: each, size: size)
        let imgView1 = UIImageView(frame: rect)
        let imgView2 = UIImageView(frame: rect)
        let index = Int(arc4random()) % images.count
        imgView1.image = images[index]
        imgView2.image = images[index]
        iv1.addSubview(imgView1)
```

```swift
            iv2.addSubview(imgView2)
        }
        //根据上面生成的3个坐标,要么上面场景中采用空白小图片,要么下面采用,由随机数的大小决定
        for (index,each) in errorCords.enumerated() {
            let size = CGSize(width: 40, height: 40)
            let rect = CGRect(origin: each, size: size)
            let btn1 = UIButton(frame: rect)
            let btn2 = UIButton(frame: rect)
            let rand = Int(arc4random() % 4)
            if rand >= 2 {
                btn1.setImage(UIImage(named: "empty.png"), for: UIControlState.normal)
                let index = Int(arc4random()) % images.count
                btn2.setImage(images[index], for: UIControlState.normal)
            }
            else {
                btn2.setImage(UIImage(named: "empty.png"), for: UIControlState.normal)
                let index = Int(arc4random()) % images.count
                btn1.setImage(images[index], for: UIControlState.normal)
            }
            btn1.isUserInteractionEnabled = true
            btn2.isUserInteractionEnabled = true
            //给按钮加上背景色,便于发现,用于提示
            btn1.backgroundColor = UIColor.blue
            btn2.backgroundColor = UIColor.green
            
            btn1.addTarget(self, action: #selector(doAction(_:)), for: UIControlEvents.touchUpInside)
            btn2.addTarget(self, action: #selector(doAction(_:)), for: UIControlEvents.touchUpInside)
            //index为数组的下标,为不同点生成不同的tag值,便于查看哪一个还没有找到
            btn1.tag = 9000 + index
            btn2.tag = 8000 + index
            
            iv1.addSubview(btn1)
            iv2.addSubview(btn2)
        }
    }
    //如果找到相应的不同点,触摸后,输出其tag值
    func doAction(_ sender : UIButton) {
        print("find\(sender.tag)")
    }
    override func didReceiveMemoryWarning() {
        super.didReceiveMemoryWarning()
        // Dispose of any resources that can be recreated.
    }
}
```

程序运行效果如图 4-4 所示。

图 4-4　找不同的运行效果

4. 选择 Main. storyboard，拖放 3 个 ViewController 并设置

从左边菜单单击 Main. storyboard 文件，在编辑区域会出现手机可视化设计界面。现在里面只有一个 View Controller 界面，从右边功能区域 ViewController，分 3 次拖放到 Main. storyboard 界面中，并分别设置 UIImageView、UILabel 和 UIButton，最后效果如图 4-5 所示。注意：在每个 View Controller 中，上面的那个 UIImageView 的 tag 设置为 1001，下面的那个 UIImageView 的 tag 设置为 1002。

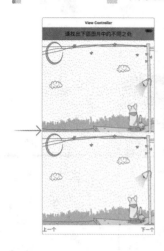   

图 4-5　4 个 View Controller 的 Main.storyboard

5. 选择 Main.storyboard，链接各个 View Controller

选中第一个 ViewController 的按钮 "下一个"，按住【Ctrl】键不放，拖放到第二个 ViewController 上，出现下拉菜单，选择其中的 Show 命令，如图 4-6 所示。

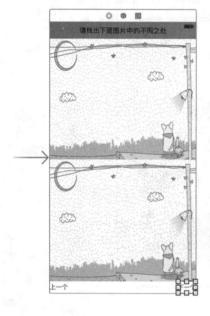 

图 4-6　在界面链接菜单选择 Show 命令

依次向下一个链接，这样就可以实现从第一个 View Controller 依次显示到最后一个 View Controller，如图 4-7 所示。如果将最后一个 "下一个" 按钮又链接到第一个 View Controller，就可以实现循环。

同样，选中最后一个 ViewController 的按钮 "上一个"，按住【Ctrl】键不放，拖放到次后一个 ViewController 上,出现下拉菜单,选择其中的 Show 命令。依次循环,实现所有"上一个"按钮的设置,这样,就可以使向前的界面循环显示,如图 4-8 所示。

图 4-7 按钮"下一个"依次链接的效果

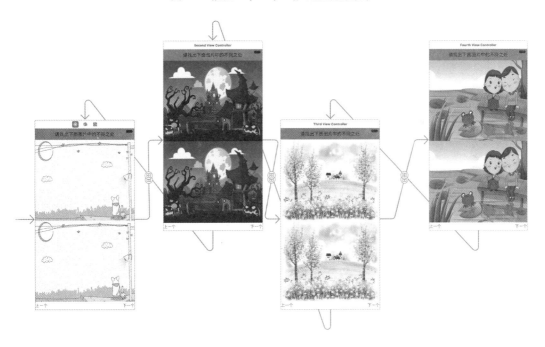

图 4-8 按钮"上一个"依次链接的效果

6. 在工程中新建 SecondViewController.swift

在工程的左侧菜单中右击,选择 New 命令,然后在对话框中选择 Cocoa Touch Class,如图 4-9 所示。

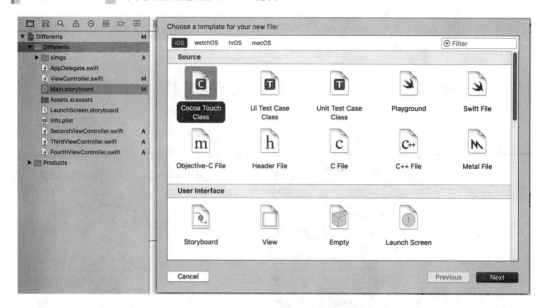

图 4-9　新建对话框中选择 Cocoa Touch Class

7. 为新建文件设置属性 Options

在 Class 中设置新建的类的名称为 SecondViewController，Subclass of 处选择 UIViewController，如图 4-10 所示。最后选择默认位置保存。

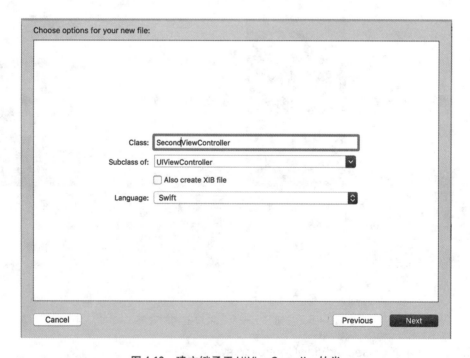

图 4-10　建立继承于 UIViewController 的类

8. 依次新建 ThirdViewController 和 FourthViewController

采用同样的方法依次新建 ThirdViewController.swift 和 FourthViewController.swift，并选

择默认位置保存。

9. 选择 Main.storyboard，链接 View Controller 和类

从左边菜单选择 Main.storyboard 文件，选择第二个 View Controller 界面，在标示查看器中的 Class 选择 SecondViewController，如图 4-11 所示。

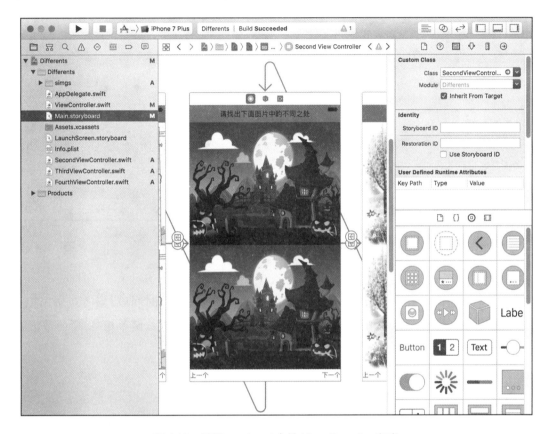

图 4-11　链接 storyboard 中的 View Controller 和类

分别实现第三个 View Controller 界面和 ThirdViewController 的链接，第四个 View Controller 界面和 FourthViewController 的链接。

10. 按照 ViewController.swift 的代码，分别输入调试

从左边菜单分别选择 SecondViewController.swift、ThirdViewController.swift 和 FourthViewController.swift 文件，根据 ViewController.swift 中代码，输入代码并调试。

4.2.2　游戏开始画面和计时功能

上面实现了游戏的基本功能，那能否为这个游戏增加一个开始界面呢？应该如何调整 Main.storyboard？实现效果如图 4-12 所示。实现方法相对比较简单，就是将一个 ViewController 拖放到 Main.storyboard 中，将这个 ViewController 设为启动项，然后添加一个按钮，将这个按钮链接到第一个 ViewController。

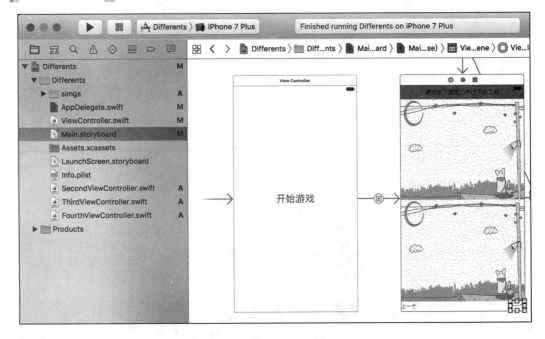

图 4-12 增加一个开始界面

如何计时功能呢？这里需要设置两个类的成员变量 createTimeStamp 和 endTime Stamp，其中 createTimeStamp 在 ViewDidAppear( )方法中获得当前时间的时间戳，当游戏玩家把 3 个不同点都找到的时候，获得当前的时间戳报错到 endTimeStamp 中，两者相减，所得数值就是玩家在本页面找不同花费的时间。

程序代码如下：

```swift
//  ViewController.swift
//  Differents
//  Created by ChenZhifeng on 17/6/4.
//  Copyright © 2017年 ChenZhifeng. All rights reserved.

import UIKit
class ViewController: UIViewController {
    //保存4个场景图片的UIImage数组
    var images : Array<UIImage> = [
        UIImage(named: "bee.png")!,
        UIImage(named: "bird.png")!,
        UIImage(named: "flower.png")!,
        UIImage(named: "mogu.png")!
    ]
    //保存各个小图片的坐标数组,上下两个场景都具有这个小图片,起到混淆作用
    var cords : Array<CGPoint> = []
    //保存不同点所在坐标的数组
    var errorCords : Array<CGPoint> = []
    //保存找到的不同点坐标的数组
    var foundIndex : Array<Int> = []
```

```swift
//保存当前场景的宽度和高度的变量Width、Height
var Width : CGFloat!
var Height : CGFloat!
//本轮开始时间,单位秒
var createTimeStamp : Int = 0
//本轮完成时间,单位秒
var endTimeStamp : Int = 0
override func viewDidLoad() {
    super.viewDidLoad()
    // Do any additional setup after loading the view, typically from a nib.
    //通过tag数值,找到上下两个场景图片的UIImageView
    let iv1 = self.view.viewWithTag(1001) as! UIImageView
    let iv2 = self.view.viewWithTag(1002) as! UIImageView
//设置可以处理触摸事件
    iv1.isUserInteractionEnabled = true
    iv2.isUserInteractionEnabled = true
    //设置小图片可以出现的最大宽度和高度,为什么都要减去40?
    Width = iv1.frame.size.width - 40
    Height = iv1.frame.size.height - 40
    //生成10个随机数坐标(x,y),保存到cords数组中
    for _ in 0...9 {
        let x = CGFloat(arc4random() % UInt32(Width))
        let y = CGFloat(arc4random() % UInt32(Height))
        cords.append(CGPoint(x: x, y: y))
    }
    //生成3个产生不同点的位置坐标,保存到errorCords数组中
    for _ in 0...2 {
        let x = CGFloat(arc4random() % UInt32(Width))
        let y = CGFloat(arc4random() % UInt32(Height))
        errorCords.append(CGPoint(x: x, y: y))
    }
    //根据上面生成的10个坐标,在上下场景中,生成一样的效果,起到混淆作用
    for each in cords {
        let size = CGSize(width: 40, height: 40)
        let rect = CGRect(origin: each, size: size)
        let imgView1 = UIImageView(frame: rect)
        let imgView2 = UIImageView(frame: rect)
        let index = Int(arc4random()) % images.count
        imgView1.image = images[index]
        imgView2.image = images[index]
        iv1.addSubview(imgView1)
        iv2.addSubview(imgView2)
    }
    //根据上面生成的3个坐标,要么上面场景中采用空白小图片,要么下面采用,由随机数的大小决定
    for (index,each) in errorCords.enumerated() {
        let size = CGSize(width: 40, height: 40)
        let rect = CGRect(origin: each, size: size)
        let btn1 = UIButton(frame: rect)
        let btn2 = UIButton(frame: rect)
        let rand = Int(arc4random() % 4)
```

```swift
                if rand >= 2 {
                    btn1.setImage(UIImage(named: "empty.png"), for: UIControlState.normal)
                    let index = Int(arc4random()) % images.count
                    btn2.setImage(images[index], for: UIControlState.normal)
                }
                else {
                    btn2.setImage(UIImage(named: "empty.png"), for: UIControlState.normal)
                    let index = Int(arc4random()) % images.count
                    btn1.setImage(images[index], for: UIControlState.normal)
                }
                btn1.isUserInteractionEnabled = true
                btn2.isUserInteractionEnabled = true
                //给按钮加上背景色,便于发现,用于提示
                btn1.backgroundColor = UIColor.blue
                btn2.backgroundColor = UIColor.green
                btn1.addTarget(self, action: #selector(doAction(_:)), for: UIControlEvents.touchUpInside)
                btn2.addTarget(self, action: #selector(doAction(_:)), for: UIControlEvents.touchUpInside)
                //index 为数组的下标,为不同点生成不同的 tag 值,便于查看哪一个还没有找到
                btn1.tag = 9000 + index
                btn2.tag = 8000 + index
                
                iv1.addSubview(btn1)
                iv2.addSubview(btn2)
            }
        }
        //如果找到相应的不同点,触摸后,输出其 tag 值
        func doAction(_ sender : UIButton) {
            var value : Int!
            if sender.tag >= 9000 {
                value = sender.tag - 9000
                for each in foundIndex {
                    if each == value {
                        return
                    }
                }
                foundIndex.append(value)
                if foundIndex.count == 3 {
                    let now = Date()
                    //当前时间的时间戳
                    let timeInterval:TimeInterval = now.timeIntervalSince1970
                    endTimeStamp = Int(timeInterval)
                    print("All found, \(endTimeStamp - createTimeStamp) Seconds")
                }
            }
            else if sender.tag >= 8000 {
                value = sender.tag - 8000
                for each in foundIndex {
                    if each == value {
```

```swift
                    return
                }
            }
            foundIndex.append(value)
            if foundIndex.count == 3 {
                let now = Date()
                //当前时间的时间戳
                let timeInterval:TimeInterval = now.timeIntervalSince1970
                endTimeStamp = Int(timeInterval)
                print("All found, \(endTimeStamp - createTimeStamp) Seconds")
            }
        }
    }
    override func viewDidAppear(_ animated: Bool) {
        let now = Date()
        //当前时间的时间戳
        let timeInterval:TimeInterval = now.timeIntervalSince1970
        createTimeStamp = Int(timeInterval)
    }
    override func didReceiveMemoryWarning() {
        super.didReceiveMemoryWarning()
        // Dispose of any resources that can be recreated.
    }

}
```

第 5 章

组建平面图形乐队

一般来说,一个乐队要有两个主唱:一男一女,应对各种歌曲;架子鼓手是整个乐队的核心之一,因为节奏手和贝斯手就是通过听架子鼓手的打拍来弹奏的;主音吉他也是乐队核心之一,主音吉他肩负着一首歌曲很重要的一部分;键盘手即电子琴手或者低频贝斯手用来打底子;节奏吉他手,配合主音吉他,完善一首歌曲。如果新组建乐队,人手不足,可以暂时取消节奏吉他手和键盘手/贝斯手,如图5-1所示。

图 5-1 演奏中的乐队

这里是模仿乐队，实现时采用多个几何图形来分别代表各个乐器或者歌手，每个几何图形和一个音乐节奏对应起来，用户可以通过触摸不同的图形来完成简单音乐的演奏。

5.1 初识图形世界

macOS 和 iOS 支持多种图形图像处理 API：UIKit、Core Graphics/Quartz 2D、Core Animation、OpenGL ES 及 Metal、Core Image、Sprite Kit/Scene Kit。这些图形 API 包含的绘制操作都是在一个图形环境中进行绘制。一个图形环境包含绘制参数和所有的绘制需要的设备特定信息，包括屏幕图形环境、位图环境和 PDF 图形环境，用来在屏幕表面、一个位图或一个 PDF 文件中进行图形和图像绘制。屏幕图形绘制限定于在一个 UIView 类或其子类的实例中，并直接在屏幕显示，也可以只在位图或 PDF 图形环境中绘制。

5.1.1 坐标系

据说有一次笛卡儿生病了，躺在床上休息，但是他的大脑却没有休息，一只在寻思通过什么手段把几何图形和代数方程关联起来，也就是几何图形中的每一个点怎么和方程的每一组解关联起来。这个时候他看到房顶上有一只蜘蛛在织网，蜘蛛空中爬来爬去。他想地上墙角的三面墙相交出三条线，把墙角作为原点，把这三条线作为数轴，那么蜘蛛某刻的位置可以通过这三条数轴上的数来表示，反过来，给定一组数便可以确定空间中的一点。后来笛卡儿发明了平面直角坐标系，当然上面的故事是三维空间的，只是为了说明，坐标系的作用是为了便于描述点的位置。

后人在笛卡儿的平面坐标系的基础上发明了三维坐标系，常用的三维坐标系分两种：左手坐标系和右手坐标系。左手坐标系和右手坐标系的规则示意如图 5-2 所示。

弯曲拇指，食指和中指使它们两两相互垂直，拇指指向 x 轴正方向，食指指向 y 轴正方向，中指指向 z 轴正方向。左手坐标系使用左手，右手坐标系使用右手。（上面示意图中的左手坐标系或者右手坐标系整体旋转后性质不变，比如左手坐标系旋转后，使得 y 轴正方向向下，x 轴正方向保持向右，它依然是左手坐标系）

另外，还有一个左手或者右手定则来判断旋转的正方向，握住拳头，拇指指向旋转轴的正方向，四指弯曲的方向为旋转的正方向。左手坐标系使用左手来判定，右手坐标系使用右手来判定，如图 5-3 图所示。

在几何图形乐队这个项目中，使用平面直角坐标系这个二维坐标系来绘制图形。

如果已经看过一些资料，Mac 和 iOS 中的各种坐标系总会让初学者摸不着头脑。不过有一点是不变的，z 轴的正方向总是指向观察者，也就是垂直屏幕平面向上。

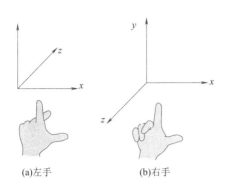
(a)左手   (b)右手

图 5-2　三维坐标系

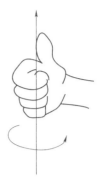

图 5-3　右手判定

1. NSView 坐标系

在 Mac 中 NSView 的坐标系默认是右手坐标系(View 其实是二维坐标系,但是为了方便可以假设其是三维坐标系,只是所有界面的变化都是在 xy 平面上),原点在左下角,如图 5-4所示。NSView 提供了一个可以用于坐标变换的方法：

- (BOOL)isFlipped;

默认返回 NO,当返回 YES 的时候,则坐标系变成左手坐标系,坐标原点变成左上角。

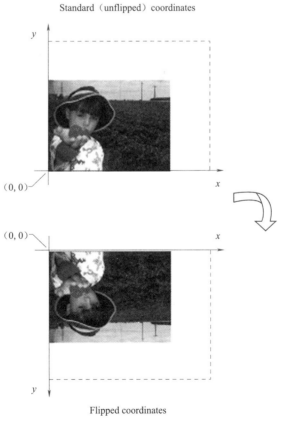

图 5-4　Mac 坐标系

在 Mac 的 AppKit 中有很多界面组件本身就使用了 Flipped Coordinate System（覆盖了上面的方法并返回 YES），如 NSButton、NSTableview、NSSplitView。

2. UIView 坐标系

而在 iOS 的 UIView 中，则没有所谓的 Flipped Coordinate 的概念，统一使用左手坐标系，也就是坐标原点在左上角，如图 5-5 所示。

Quartz（Core Graphics）坐标系使用的右手坐标系，原点在左下角，所以所有使用 Core Graphics 画图的坐标系都是右手坐标系。当使用 Core Graphics 的相关函数画图到 UIView 上的时候，需要注意 CTM 的 Flip 变换，要不然会出现界面上图形倒过来的现象。由于 UIKit 提供的高层方法会自动处理 CTM（通信终端模块，比如 UIImage 的 drawInRect( )方法），所以无须自己在 CG 的上下文中进行处理。

CALayer 坐标系，其坐标系和平台有关，在 Mac 中 CALayer 使用的是右手坐标系，其原点在左下角；iOS 中使用的左手坐标系，其原点在左上角。

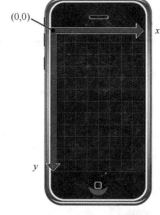

图 5-5　iOS 采用左上角坐标系

5.1.2　UIKit 和 Core Graphics

UIKit 绘图方式实际是对 Core Graphics 方式的一种简化封装，可以采用面向对象的方式很方便地做各种绘图操作，主要是通过 UIBezierPath 这个类来实现的，创建基于矢量的路径，例如各种直线、曲线、圆等等。

Core Graphics 是基于 C 语言的 API，可以用于一切 iOS 绘图程序的开发，如图 5-6 所示。Quartz 2D 是 Core Graphics 框架的一部分，是一个强大的二维图像绘制引擎。Quartz 2D 在 UIKit 中也有很好的封装和集成，日常所用到的 UIKit 中的各种组件都是由 Core Graphics 进行绘制的，包括一些常用的绘图 API。图形上下文 CGContext 代表图形输出设备（也就是绘制的位置），包含了绘制图形的一些设备信息。Quartz 2D 中所有对象都必须绘制在图形上下文。

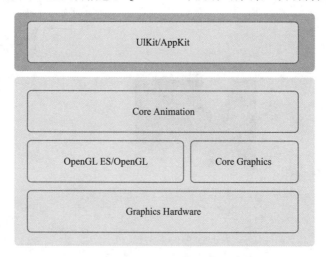

图 5-6　Core Graphics 在绘图 API 中的层次

Core Graphics 绘图的一般步骤，如图 5-7 所示，主要包括：

（1）获取图形上下文。

（2）创建并设置路径，将路径添加到上下文，如直线的坐标等。

（3）设置上下文状态，如颜色、线条宽度等。

（4）绘制路径。

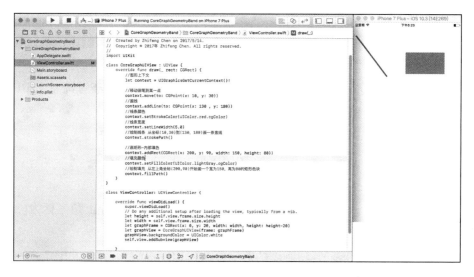

图 5-7　Core Graphics 绘图的示例

Quartz 2D 路径可以用来描述矩形、圆，以及其他想要的 2D 几何图形。通过路径可以对几何图形进行各种处理。Quartz 2D 中有 4 种基本图元：点、线段、弧和贝塞尔曲线。

点：点是二维空间中的一个位置，不等同于像素，一个点完全不占空间。画一个点不会在屏幕上显示任何东西。

线段：线段由起点和终点两个点定义。线段没有面积，所以它们不能被填充。可以用一组线段或者曲线组成一个具有闭合路径的几何图形，然后进行填充。

弧：弧由一个圆心点、半径、起始角和结束角定义。圆是弧的特例，弧是占有一定面积的路径，所以可以被填充、描边和描边填充出来。

贝塞尔曲线：任何一条曲线都可以通过与它相切的控制线两端的点的位置来定义。贝塞尔曲线可以用 4 个点描述，其中两个点描述两个端点，另外两个描述每一端的切线。

在 iOS 上无论采用哪种绘图技术，绘制都发生在 UIView 对象的区域内，因此，可以重载 UIView 的 draw(_ rect：CGRect)方法中实现自定义绘图。

系统会为视图设置一个重绘标识，在每次循环时，绘图引擎会检查重绘标识，以此判断是否有需要更新的内容，如果需要重绘，则自动调用 draw(_ rect：CGRect)方法。

当然，也可以手动设置重绘标识，一般可以调用以下函数来实现重绘：

（1）重绘整个视图：setNeedsDisplay( )。

（2）重绘指定区域的视图：setNeedsDisplay（rect：CGRect）。

原则上尽量不要重复绘制全部视图，以降低系统绘制开销，以下几种情况会触发视图重绘：

（1）当遮挡视图的其他视图被移动或删除操作的时候。

（2）将视图的 hidden 属性声明设置为 NO，使其从隐藏状态变为可见。

（3）将视图滚出屏幕，然后再重新回到屏幕上。

（4）显式调用视图的 setNeedsDisplay 或者 setNeedsDisplay（rect：CGRect）方法。

下面就以一个实例来分析一下具体的步骤，体会一下 Quartz 2D 的绘图效果。在 Quartz 2D 的绘图中常见的函数及其功能如下：

（1）draw（_ rect：CGRect）函数：这个是 UIView 内部自动调用，用于重画整个 UIView 中的图形内容，一般需要重载，实现自定义绘图。

（2）UIGraphicsGetCurrentContext（）函数：用于获得当前的图形上下文。

（3）UIGraphicsEndImageContext（）函数：关闭图形上下文。

（4）CGPoint（x：，y：）函数：建立一个点，其坐标为（x,y）。

（5）CGRect（x：，y：，width：，height：）函数：建立一个矩形，其左上角坐标为（x,y），其宽为 width，其高为 height。

（6）move（to：CGPoint）函数：将画图的落笔移动到 CGPoint 这一点所在的位置，开始画。

（7）addLine（to：CGPoint）函数：从当前点开始，画一条线段，其终点为 CGPoint 所在位置。

（8）setStrokeColor（CGColor）函数：设置笔画线段的颜色。

（9）setFillColor（CGColor）函数：设置填充的颜色。

（10）setLineWidth（CGFloat）函数：设置笔画线段的粗细。

（11）addRect（CGRect）函数：根据 CGRect 的矩形位置和大小，画出一个矩形。

（12）strokeEllipse（in：CGRect）函数：根据 CGRect 矩形的大小，画出了椭圆形。

（13）addArc（center：，radius：，startAngle：，endAngle：，clockwise：）函数：以 center 为中心，半径为 radius，开始位置为 startAngle，圆弧结束位置为 endAngle（以弧度表示，CGFloat.pi＊2 代表整个圆），clockwise 为 true 代表为顺时针方向。

（14）strokePath（）函数：立即根据前面设置的路径画线。

（15）fillPath（）函数：立即根据前面设置的路径填充颜色。

（16）closePath（）函数：从当前的位置，连接到起始点所在的位置。

（17）context.setLineDash（phase：0，lengths：[6,1]）函数：设置为虚线，其中 phase 为 0 代表从头开始，lengths 为[6,1]代表实线和虚线的比例为 6:1。

（18）drawPath（using：CGPathDrawingMode.fillStroke）函数：可以实现笔画线段和内部填充同时进行，相当于 strokePath（）和 fillPath（）的功能。

（19）CoreGraphUIView（）：从 UIView 继承而来的一个自定义的类，在这个新类中，重载了draw(_ rect：CGRect)函数，从而实现绘图功能。

以下为直线、三角形和矩形其交的绘制代码：

```swift
// ViewController.swift
// CoreGraphGeometryBand
// Created by Zhifeng Chen on 2017/5/14.
// Copyright ? 2017 年 Zhifeng Chen. All rights reserved.
import UIKit
class CoreGraphUIView : UIView {
    override func draw(_ rect: CGRect) {
        //图形上下文
        let context = UIGraphicsGetCurrentContext()!
        
        //移动画笔到某一点
        context.move(to: CGPoint(x: 350, y: 30))
        //画线
        context.addLine(to: CGPoint(x: 200 , y: 85))
        //线条颜色
        context.setStrokeColor(UIColor.red.cgColor)
        //线条宽度
        context.setLineWidth(5.0)
        //绘制线条 从坐标(350,30)到(200,85)画一条直线
        context.strokePath()
        
        //画矩形 - 内部填色
        context.addRect(CGRect(x: 200, y: 90, width: 150, height: 80))
        //填充颜色
        context.setFillColor(UIColor.lightGray.cgColor)
        //绘制填充 从左上角坐标(200,90)开始画一个宽为150,高为80 的矩形色块
        context.fillPath()
        
        //设置三角形的3个顶点坐标
        context.move(to: CGPoint(x: 95, y: 10))
        context.addLine(to: CGPoint(x: 40, y: 150))
        context.addLine(to: CGPoint(x: 160,y: 190))
        //闭合路径
        context.closePath()
        //线条颜色
        context.setStrokeColor(UIColor.red.cgColor)
        //绘制线条 从坐标(95,10),到坐标(40,150),到坐标(160,190),
        //最后回到起点(95,10)
        context.strokePath()
    }
}
class ViewController: UIViewController {
    override func viewDidLoad() {
        super.viewDidLoad()
        // Do any additional setup after loading the view
        let height = self.view.frame.size.height
        let width = self.view.frame.size.width
```

```
            let graphFrame = CGRect(x: 0, y: 0, width: width, height: height)
            let graphView = CoreGraphUIView(frame: graphFrame)
            graphView.backgroundColor = UIColor.white
            self.view.addSubview(graphView)
    }
}
```

程序运行后的效果如图 5-8 所示。

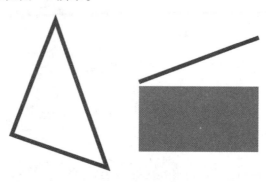

图 5-8 直线、三角形和矩形填充的绘制

在 Quartz 2D 中曲线绘制分为两种：二次贝塞尔曲线和三次贝塞尔曲线。二次曲线只有一个控制点，而三次曲线有两个控制点，如图 5-9 所示。

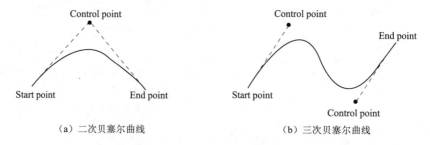

（a）二次贝塞尔曲线　　　　　　　　（b）三次贝塞尔曲线

图 5-9 贝塞尔曲线

Core Graphics 不仅可以画图还能绘制文字，还能将图像文件绘制出来，如图 5-10 所示。

图 5-10 二次贝塞尔曲线、文字和图片的绘制

详细的源代码如下：

```swift
// ViewController.swift
// CoreGraphGeometryBand
// Created by Zhifeng Chen on 2017/5/14.
// Copyright ? 2017年 Zhifeng Chen. All rights reserved.
import UIKit
class CoreGraphUIView : UIView {
    override func draw(_ rect: CGRect) {
        //图形上下文
        let context = UIGraphicsGetCurrentContext()!
        
        //二次贝塞尔曲线的绘制
        //创建并设置路径
        let path = CGMutablePath()
        //虚线
        context.setLineDash(phase: 0, lengths: [6,5])
        //移动起点
        path.move(to: CGPoint(x: 10, y: 300))
        //二次贝塞尔曲线终点
        let toPoint = CGPoint(x: 200, y: 300)
        //二次贝塞尔曲线控制点
        let controlPoint = CGPoint(x: 135, y: 60)
        //绘制二次贝塞尔曲线
        path.addQuadCurve(to: toPoint, control: controlPoint)
        //添加路径到图形上下文
        context.addPath(path)
        //设置笔划颜色
        context.setStrokeColor(UIColor.orange.cgColor)
        //设置填充颜色
        context.setFillColor(UIColor.lightGray.cgColor)
        //设置笔触宽度
        context.setLineWidth(6)
        //绘制路径
        context.drawPath(using: CGPathDrawingMode.fillStroke)
        
        //文字的绘制
        let s : String = "二次贝塞尔曲线的绘制"
        (s as NSString).draw(at: CGPoint(x: 30, y: 120), withAttributes: [NSFontAttributeName:UIFont.boldSystemFont(ofSize: 28),NSForegroundColorAttributeName:UIColor.orange])
        
        //图片的绘制
        let img = UIImage(named: "bee.png")
        //绘制到指定的矩形中,注意如果大小不合适会进行拉伸
        img?.draw(in: CGRect(x:250, y:200, width:150, height:180))
        //从某一点开始绘制
        img?.draw(at: CGPoint(x: 0, y: 320))
    }
}
class ViewController: UIViewController{
    override func viewDidLoad(){
        super.viewDidLoad()
```

```
            // Do any additional setup after loading the view,
            //typically from a nib.
            let height = self.view.frame.size.height
            let width = self.view.frame.size.width
            let graphFrame = CGRect(x: 0, y: 20, width: width, height: height-20)
            let graphView = CoreGraphUIView(frame: graphFrame)
            graphView.backgroundColor = UIColor.white
            self.view.addSubview(graphView)
        }
    }
```

5.2 触摸事件和绘图

UIView 就是表示屏幕上的一块矩形区域,它在 APP 中占有绝对重要的地位,因为 iOS 中几乎所有的可视控件都是 UIView 的子类。

UIView 继承自 UIResponder,它是负责显示的画布,如果说把 window 比作画框,我们就是不断地在画框上移除、更换或者叠加画布,或者在画布上叠加其他画布,大小当然由绘画者来决定。有了画布,就可以在上面任意施为。UIView 的功能包括:管理视图区域中的内容、处理视图区域中的事件、管理子视图,以及绘图、动画等,如图 5-11 所示。

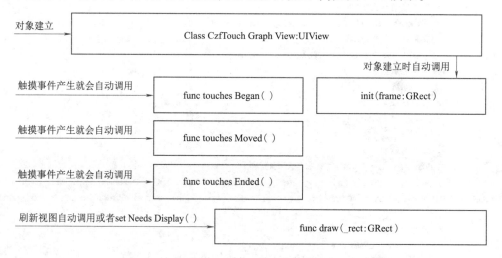

图 5-11　UIView 中各个事件的调用

UIView 的绘图方法和触摸事件,其中触摸事件继承自 UIResponder 的触摸事件的方法:

(1) func draw(_ rect:CGRect):重载该函数用于自定义绘图(UIView)。

(2) func draw(_ layer：CALayer, in ctx：CGContext):重载该函数用于自定义绘图(CALayer)。

(3) func touchesBegan:withEvent:当用户触摸到屏幕时调用此方法。

（4）func touchesEnded:withEvent：当触摸离开屏幕时调用此方法。

（5）func touchesMoved:withEvent：当用户触摸到屏幕并移动时调用此方法。

（6）func touchesCancled:withEvent：当触摸被取消时调用此方法。

在用户使用 APP 过程中，会产生各种各样的事件，iOS 中的事件可以分为三大类型，如图 5-12 所示。

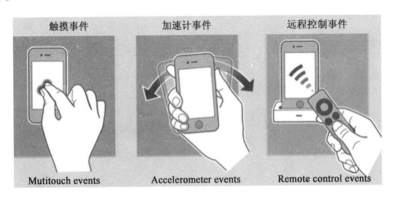

图 5-12　iOS 中的三类事件

在 iOS 中不是任何对象都能处理事件，只有继承了 UIResponder 的对象才能接收并处理事件，通常称之为"响应者对象"。UIApplication、UIViewController、UIView 都继承自 UIResponder，因此它们都是响应者对象，都能够接收并处理事件，如图 5-13 所示。

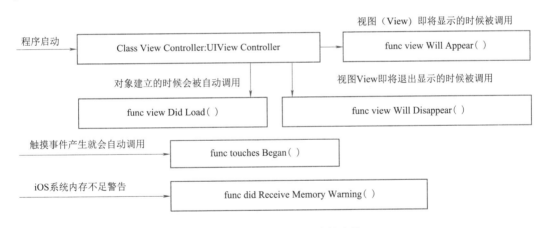

图 5-13　UIViewController 中的事件

一次完整的触摸过程中，只会产生一个事件对象，4 个触摸方法都是同一个 event 参数，如果两个手指同时触摸一个视图，那么 view 只会调用一次 touchesBegan:withEvent 方法，touches 参数中装着 2 个 UITouch 对象；如果这两个手指一前一后分开触摸同一个视图，那么会分别调用 2 次 touchesBegan:withEvent 方法，并且每次调用时的 touches 参数中只包含一个 UITouch 对象。

【例 5-1】在 UIView 上触摸、移动和停止触摸会引起屏幕的颜色变化，可以从中体会 touchesBegan、touchesEnded 和 touchesMoved 三个触摸事件的使用。

程序代码如下：

```swift
// ViewController.swift
// CoreGraphGeometryBand
// Created by Zhifeng Chen on 2017/5/14.
// Copyright ? 2017 年 Zhifeng Chen. All rights reserved.
//
import UIKit
class CoreGraphUIView : UIView {
    override func touchesBegan(_ touches: Set<UITouch>, with event: UIEvent?) {
        self.backgroundColor = UIColor.red
        print("Began: \(touches)")
    }
    override func touchesEnded(_ touches: Set<UITouch>, with event: UIEvent?) {
        self.backgroundColor = UIColor.lightGray
        print("Ended: \(touches)")
    }
    override func touchesMoved(_ touches: Set<UITouch>, with event: UIEvent?) {
        self.backgroundColor = UIColor.blue
        print("Moved: \(touches)")
    }
}
class ViewController: UIViewController {
    override func viewDidLoad() {
        super.viewDidLoad()
        // Do any additional setup after loading the view
        let height = self.view.frame.size.height
        let width = self.view.frame.size.width
        let graphFrame = CGRect(x: 0, y: 0, width: width, height: height)
        let graphView = CoreGraphUIView(frame: graphFrame)
        graphView.backgroundColor = UIColor.white
        self.view.addSubview(graphView)
    }
}
```

如果想获得触摸的对象及其坐标，程序如下：

```swift
let touch:UITouch = touches.first! as UITouch
print(touch.location(in: view).x)
print(touch.location(in: view).y)
```

【例5-2】在用户触摸后，会在屏幕上生成大大小小的正方形颜色块，主要用到了touchesBegan()触摸事件。

程序代码如下：

```swift
// ViewController.swift
// CirecleTouchView
// Created by Zhifeng Chen on 2017/5/15.
// Copyright ? 2017 年 Zhifeng Chen. All rights reserved.
import UIKit
```

```
class ViewController: UIViewController {
    //建立一个UIView的变量
    var myView : UIView!
    //重载触摸事件中的touchesBegan
    override func touchesBegan(_ touches: Set<UITouch>, with event: UIEvent?) {
        // 获得UITouch集合
        let touch:UITouch = touches.first! as UITouch
        // 获得触摸所在位置的坐标
        let center = touch.location(in: view)
        // 采用随机数函数生成矩形的宽width
        let width = CGFloat(25 + (arc4random() % 50))
        // 高度等于宽度
        let height = width
        // 根据矩形框的来生成UIView的对象,设置触摸点所在位置为矩形的中心点
        myView = UIView(frame: CGRect(x:center.x - width/2 , y:center.y - height/2, width:width, height:height))
        // 设置背景色为蓝色
        myView.backgroundColor = UIColor.blue
        // 将myView添加到主界面中
        view.addSubview(myView)
    }
    override func viewDidLoad() {
        super.viewDidLoad()
        // Do any additional setup after loading the view, typically from a nib.
    }
}
```

程序运行效果如图5-14所示。

练习:如果知道一个矩形的中心坐标(center.x,center.y),以及宽(let width = frame.size.width)和高(let height = frame.size.height),如果计算给矩形的左上角坐标?

提示:该矩形的左上角坐标就是(center.x – width/2, center.y – height/2),如图5-15所示。

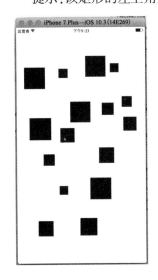

图5-14 随机大小的方块

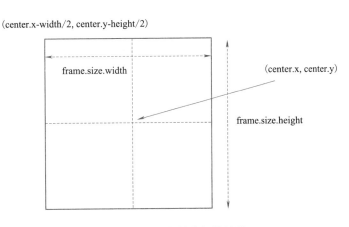

图5-15 矩形相关坐标的计算

【例 5-3】 在上述程序的基础上进行修改，会在屏幕上生成大大小小的方块。方块在生成的时候随触摸移动事件而移动，在触摸结束的时候，会将方块放置在该位置，同时将其设置为另外一种颜色。本程序演示了触摸事件的 touchesBegan( )、touchesMoved( ) 和 touchesEnded( ) 使用方法。

程序代码如下：

```swift
// ViewController.swift
// CirecleTouchView
// Created by Zhifeng Chen on 2017/5/15.
// Copyright ? 2017 年 Zhifeng Chen. All rights reserved.
import UIKit
class ViewController: UIViewController {
    //建立一个UIView的变量
    var myView : UIView!
    override func touchesBegan(_ touches: Set<UITouch>, with event: UIEvent?) {
        // 获得UITouch集合
        let touch:UITouch = touches.first! as UITouch
        // 获得触摸所在位置的坐标
        let center = touch.location(in: view)
        // 采用随机数函数生成矩形的宽width
        let width = CGFloat(25 + (arc4random()%50))
        // 高度等于宽度
        let height = width
        // 根据矩形框的来生成UIView的对象,设置触摸点所在位置为矩形的中心点
        myView = UIView(frame: CGRect(x:center.x - width/2 , y:center.y - height/2, width:width, height:height))
        // 设置背景色blue
        myView.backgroundColor = UIColor.blue

        // 将myView添加到主界面中
        view.addSubview(myView)
    }
    override func touchesMoved(_ touches: Set<UITouch>, with event: UIEvent?) {
        // 获得UITouch集合
        let touch:UITouch = touches.first! as UITouch
        // 获得触摸所在位置的坐标
        let center = touch.location(in: view)
        // 获得myView的width
        let width = myView.frame.width
        // 获得myView的height
        let height = myView.frame.height
        // 重新计算新的位置
        let newRect = CGRect(x:center.x - width/2 , y:center.y - height/2, width:width, height:height)
        // 设置myView到新的位置
        myView.frame = newRect
    }
    override func touchesEnded(_ touches: Set<UITouch>, with event: UIEvent?) {
        //将最终的设置颜色为orange
```

```
        myView.backgroundColor = UIColor.orange
    }
    override func viewDidLoad() {
        super.viewDidLoad()
        // Do any additional setup after loading the view, typically from a nib.
    }
}
```

5.3 拥有自己的绘图类

Swift 提供了全面的面向对象支持,与普通面向对象的编程语言的不同之处在于,不仅可以定义类,还支持枚举和结构体、扩展和协议的面向对象支持。其中,枚举和结构体是值类型,而类定义的变量则是引用类型。

面向对象编程是以对象为中心的编程方式,其三大典型特征如下:

(1)封装:指的是把对象的状态数据、实现细节隐藏起来,然后暴露合适的方法允许外部程序改变对象的状态。Swift 提供了 private、internal 和 public 等访问权限控制。

(2)继承:指的是子类继承父类,即可获得父类定义的属性和方法,通过继承可以复用已有类的方法和属性。Swift 提供了很好的单继承:每个子类最多只能有一个直接父类。另外,协议可以有效弥补单继承的不足。

(3)多态:指的是利用面向对象的灵活性,使得同名函数可以实现不同的功能。

5.3.1 继承 UIView 建立自己的图形类

用户要在 iOS 上画图,需要继承 UIView,然后重载里面的 draw(_ rect: CGRect)方法,就可以自定义画图。draw(_ rect: CGRect)函数是由 UIView 内部自动调用或者通过 setNeedsDisplay()函数来调用。因此,需要重载 draw(_ rect: CGRect)函数,把需要绘图的程序编写在里面。

【例 5-4】在 UIView 上触摸会在触摸所在的位置生成一个圆,每个圆的大小是随机的,圆心的位置就是触摸所在的坐标点,效果如图 5-16 所示。主要重载了 touchesBegan 事件,同时自定义了一个 CircleView 类(基类为 UIView),重载了其中的 draw(_ rect: CGRect),以及重载了 init(frame: CGRect)初始化方法。

程序代码如下:

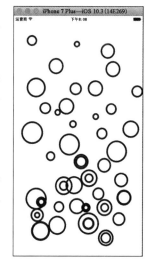

图 5-16 随机大小的圆圈

```
// ViewController.swift
// CircleCreate
```

```swift
//  Created by Zhifeng Chen on 2017/6/6.
//  Copyright ? 2017 年 Zhifeng Chen. All rights reserved.
import UIKit
class CircleView : UIView {
    //重载类初始化函数,设置背景色清除
    override init(frame: CGRect) {
        super.init(frame: frame)
        self.backgroundColor = UIColor.clear
    }
    required init?(coder aDecoder: NSCoder) {
        fatalError("init(coder:) has not been implemented")
    }
    override func draw(_ rect: CGRect) {
        //调用 UIBezierPath()建立路径绘图
        let path = UIBezierPath()
        //生成一个圆弧,从 0 到 2pi,圆心在 View 的中心点,半径为 View 高度的一半 -5,顺时针方向
        path.addArc(withCenter: CGPoint(x:self.frame.size.width/2, y:self.frame.size.height/2), radius: self.frame.size.height/2 - 5, startAngle: 0, endAngle: CGFloat.pi * 2, clockwise: true)
        //圆弧的内部填充色为绿色
        UIColor.green.setFill()
        //填充圆弧内部
        path.fill()
        //设置线宽为 5
        path.lineWidth = 5
        //设置线的颜色为红色
        UIColor.red.setStroke()
        //画圆弧轮廓线
        path.stroke()
    }
}
class ViewController: UIViewController {
    //建立一个 UIView 的变量 -> 建立一个 CircleView 的变量
    var myView : CircleView!
    override func touchesBegan(_ touches: Set<UITouch>, with event: UIEvent?) {
        // 获得 UITouch 集合
        let touch:UITouch = touches.first! as UITouch
        // 获得触摸所在位置的坐标
        let center = touch.location(in: view)
        // 采用随机数函数生成矩形的宽 width
        let width = CGFloat(25 + (arc4random()%50))
        // 高度等于宽度
        let height = width
        // 根据矩形框的来生成 UIView 的对象,设置触摸点所在位置为矩形的中心点
        // 将 UIView 修改为 CircleTouchesBeganView
        myView = CircleView(frame: CGRect(x:center.x - width/2 , y:center.y - height/2, width:width, height:height))
        // 将 myView 添加到主界面中
        view.addSubview(myView)
    }
    override func viewDidLoad() {
```

```
        super.viewDidLoad()
        // Do any additional setup after loading the view, typically from a nib.
    }
    override func didReceiveMemoryWarning(){
        super.didReceiveMemoryWarning()
        // Dispose of any resources that can be recreated.
    }
}
```

【例5-5】在UIView上触摸会在触摸所在的位置生成一个圆,每个圆的大小是随机的,圆心的位置就是触摸所在的坐标点,如果继续移动,这个圆会跟随着你移动,直到停止触摸,就会固定在那个位置,同时改变颜色。主要重载了touchesBegan、touchesMoved和touchesEnded事件,同时在ViewController中定义了一个CircleView类(基类为UIView)的变量myCircle,使用了setNeedsDisplay()进行图形刷新。

程序代码如下:

```
// ViewController.swift
// CircleCreate
// Created by Zhifeng Chen on 2017/6/6.
// Copyright ? 2017年 Zhifeng Chen. All rights reserved.
import UIKit
class CircleView : UIView {
    //保存当前线条的颜色,默认值为blue
    var lineColor : UIColor = UIColor.blue
    //重载类初始化函数,设置背景色清除
    override init(frame: CGRect) {
        super.init(frame: frame)
        self.backgroundColor = UIColor.clear
    }
    required init? (coder aDecoder: NSCoder) {
        fatalError("init(coder:) has not been implemented")
    }
    override func draw(_ rect: CGRect) {
        //调用UIBezierPath()建立路径绘图let path = UIBezierPath()
        //生成一个圆弧,从0到2pi,圆心在View的中心点,半径为View高度的一半-5,顺时针方向
        path.addArc(withCenter: CGPoint(x:self.frame.size.width/2,y:self.frame.size.height/2), radius: self.frame.size.height/2 - 5, startAngle: 0, endAngle: CGFloat.pi * 2, clockwise: true)
        //圆弧的内部填充色为绿色
        UIColor.green.setFill()
        //填充圆弧内部
        path.fill()
        //设置线宽为5
        path.lineWidth = 5
        //设置线的颜色为红色
        lineColor.setStroke()
        //画圆弧轮廓线
        path.stroke()
    }
}
```

```swift
class ViewController: UIViewController {
    //建立一个UIView的变量 -> 建立一个CircleView的变量
    var myView : CircleView!
    override func touchesBegan(_ touches: Set<UITouch>, with event: UIEvent?) {
        // 获得UITouch集合
        let touch:UITouch = touches.first! as UITouch
        // 获得触摸所在位置的坐标
        let center = touch.location(in: view)
        // 采用随机数函数生成矩形的宽width
        let width = CGFloat(25 + (arc4random() % 50))
        // 高度等于宽度
        let height = width
        // 根据矩形框的来生成UIView的对象,设置触摸点所在位置为矩形的中心点
        // 将UIView修改为CircleTouchesBeganView
        myView = CircleView(frame: CGRect(x:center.x - width/2 , y:center.y - height/2, width:width, height:height))
        // 将myView添加到主界面中
        view.addSubview(myView)
    }
    override func touchesMoved(_ touches: Set<UITouch>, with event: UIEvent?) {
        // 获得UITouch集合
        let touch:UITouch = touches.first! as UITouch
        // 获得触摸所在位置的坐标
        let center = touch.location(in: view)
        // 获得myView的width
        let width = myView.frame.width
        // 获得myView的height
        let height = myView.frame.height
        // 重新计算新的位置
        let newRect = CGRect(x:center.x - width/2 , y:center.y - height/2, width:width, height:height)
        // 设置myView到新的位置
        myView.frame = newRect
    }
    override func touchesEnded(_ touches: Set<UITouch>, with event: UIEvent?) {
        //将最终的设置颜色为orange
        myView.lineColor = UIColor.orange
        //要求myView立即调用draw(),颜色被刷新
        myView.setNeedsDisplay()
    }
    override func viewDidLoad() {
        super.viewDidLoad()
        // Do any additional setup after loading the view, typically from a nib.
    }
        override func didReceiveMemoryWarning() {
        super.didReceiveMemoryWarning()
        // Dispose of any resources that can be recreated.
    }
}
```

那么，能否在拖动的时候也改变颜色呢？这就只需要重载 touchesMoved( )函数,修改其中的代码即可。

程序代码如下：

```swift
override func touchesMoved(_ touches: Set<UITouch>, with event: UIEvent?) {
    let colors : Array<UIColor> = [
        UIColor.cyan,UIColor.brown,UIColor.cyan,
        UIColor.gray,UIColor.lightGray
    ]
    // 获得 UITouch 集合
    let touch:UITouch = touches.first! as UITouch
    // 获得触摸所在位置的坐标
    let center = touch.location(in: view)
    // 获得 myView 的 width
    let width = myView.frame.width
    // 获得 myView 的 height
    let height = myView.frame.height
    // 重新计算新的位置
    let newRect = CGRect(x:center.x - width/2 , y:center.y - height/2    , width:width, height:height)
    // 设置 myView 到新的位置
    myView.frame = newRect
    //根据鼠标位置坐标(x,y)的整数值,获得在数组的下标
    let index = Int(center.x + center.y) % colors.count
    //颜色赋值
    myView.lineColor = colors[index]
    //要求 myView 立即调用 draw(),颜色被刷新
    myView.setNeedsDisplay()
}
```

习题：如果要生成其他图形该如何处理程序,请修改程序,实现如图 5-17 的图形。

图 5-17　要生成的图形

程序代码如下：

```swift
class HalfCircleView : UIView {
    //保存当前线条的颜色,默认值为 blue
    var lineColor : UIColor = UIColor.blue
    //重载类初始化函数,设置背景色清除
    override init(frame: CGRect) {
        super.init(frame: frame)
        self.backgroundColor = UIColor.clear
    }
```

```swift
        required init? (coder aDecoder: NSCoder) {
            fatalError("init(coder:) has not been implemented")
        }
        override func draw(_ rect: CGRect) {
            //调用UIBezierPath()建立路径绘图
            let path = UIBezierPath()
            //生成一个圆弧,从0到2pi,圆心在View的中心点,半径为View高度的一半-5,顺时针方向
            path.addArc(withCenter: CGPoint (x:self.frame.size.width/2, y:self.frame.size.height/2), radius: self.frame.size.height/2 - 5, startAngle: 0, endAngle: CGFloat.pi, clockwise: false)
            // 生成三条线段
            path.move(to: CGPoint(x: 0, y: frame.size.height/2))
            path.addLine(to: CGPoint(x: 0, y: frame.size.height))
            path.addLine(to: CGPoint(x: frame.size.width, y: frame.size.height))
            path.addLine(to: CGPoint(x: frame.size.width, y: frame.size.height/2))
            //圆弧的内部填充色为绿色
            UIColor.green.setFill()
            //填充圆弧内部
            path.fill()
            //设置线宽为5
            path.lineWidth = 5
            //设置线的颜色为红色
            lineColor.setStroke()
            //画圆弧轮廓线
            path.stroke()
        }
    }
```

5.3.2 从无到有建立自己的图形类

类一般采用 class 来声明。对于一个类定义而言,可以包含4种最常见的成员:构造器、属性、方法、下标。各个成员之间的定义顺序没有影响,成员之间可以相互调用。

属性用于定义该类或该类的实例所包含的状态数据;方法则用于定义该类或该类的实例的行为特征或者功能实现;构造器用于构造该类的实例。

构造器是一个类创建实例的重要途径,如果一个类没有构造器,这个类就无法创建实例。因此,Swift 提供了一个功能:如果程序员没有为一个类提供构造器,则系统会为该类提供一个默认的、无参数的构造器。一旦程序员为该类提供了构造器,系统将不再为该类提供默认的、无参数的构造器。

修饰符 class 和 static 作用基本相同:没有 class 或 static 修饰的是实例成员,有 class 或 static 修饰的是类型成员。

注意:采用 class 或者 static 修饰的成员(类成员)不能访问没有 class 或 static 修饰的成员(实例成员)。

Swift 的属性分为两种:存储属性和计算属性。存储属性用于保存类型本身或实例的状

态数据;计算属性则采用 setter、getter 方法合成,不一定保存状态数据。

类的方法从语法格式上看和函数完全一致。构造器就是一个特殊的方法,其名称固定为 init,且不能声明返回值类型,实际上 Swift 构造器隐式地返回 self 作为返回值。

1. 定义一个 Shape 类

下面程序定义一个 Shape 类,实现用于描述图形的一个基类:包括存储属性 6 个、构造器 1 个、方法 1 个。

程序代码如下:

```
class Shape {
    //名称
    var name : String?
    //边数
    var sides : Int?
    //左上角的位置坐标
    var origin : CGPoint?
    //线条颜色
    var lineColor : UIColor = UIColor.red
    //填充颜色
    var fillColor : UIColor = UIColor.green
    //线条宽度
    var lineWidth : CGFloat = 2
    //构造器函数 init
    init(name : String, sides : Int, origin : CGPoint) {
        self.name = name
        self.sides = sides
        self.origin = origin
    }
    //自定义方法 sayHello
    func sayHello(){
        print("Shape is \(name!),sides \(sides!), and originCord is(\(origin!.x),\(origin!.y)) ")
    }
}
```

2. 实例化 Shape 类

类设计完成后,如何来使用这个类? 这就需要进行实例化。下面新建一个工程 classDemo,来实例化这个类。

程序代码如下:

```
// ViewController.swift
// classDemo
// Created by Zhifeng Chen on 2017/3/4.
// Copyright ? 2017 年 Zhifeng Chen. All rights reserved.

import UIKit
class Shape {
    //名称
    var name : String?
```

```swift
        //边数
        var sides : Int?
        //左上角的位置坐标
        var origin : CGPoint?
        //线条颜色
        var lineColor : UIColor = UIColor.red
        //填充颜色
        var fillColor : UIColor = UIColor.green
        //线条宽度
        var lineWidth : CGFloat = 2
        //构造器函数 init
        init(name : String, sides : Int, origin : CGPoint) {
            self.name = name
            self.sides = sides
            self.origin = origin
        }
        //自定义方法 sayHello
        func sayHello(){
            print("Shape is \(name!),sides \(sides!), and originCord is(\(origin!.x),\(origin!.y)) ")
        }
}
class ViewController: UIViewController {
    override func viewDidLoad(){
        super.viewDidLoad()
        // Do any additional setup after loading the view, typically from a nib.
        //此处调用 Shape 类,建立一个对象(实例)myShaple
        let myShape = Shape(name:"BaseShape", sides: 0, origin: CGPoint(x: 0, y: 0))
        myShape.sayHello()
    }
    override func didReceiveMemoryWarning(){
        super.didReceiveMemoryWarning()
        // Dispose of any resources that can be recreated.
    }
}
```

程序启动时,会自动调用 ViewController 中的 ViewDidLoad( )方法,从而实现 Shape 类的实例,其实例化的对象名称(变量)为 myShape。最终结果是在控制台输出"Shape is BaseShape,sides 0, and originCord is(0,0)"。

类的实例化非常简单,一般就是通过调用类的构造器"类名(参数)"来实现。

大部分时候,定义一个类是为了重复创建该类的实例,同一个类的多个实例具有相同的特征,而类则是定义了多个实例的共同特征。从某个角度来看,类定义的是多个实例的一种抽象特征,实例才是具体存在,拥有分配的内存空间。

注意:类是引用类型,而常见的其他类型如枚举、结构体等属于值类型。所谓引用类型可以理解为 C 语言中指针类型。

Swift 为每个类提供了一个 self 关键字,self 总是指向该方法的调用者。在实例方法中,

self 代表调用该方法的实例;在类型方法中,self 代表调用给方法的类本身。

3. 重载 UIView 调用 Shape 类

【例5-6】通过在 ViewController 的 ViewDidLoad( )方法中分别实例化 Shape 类和 CzfView 类,从而实现在 CzfView 类重载的方法 draw( )中,调用 Shape 类中的方法 sayHello( )。通过观察控制台信息输出,可以看到 sayHello( )的输出信息。

程序代码如下:

```swift
// ViewController.swift
// DrawMusic
// Created by Zhifeng Chen on 2017/6/4.
// Copyright ? 2017年 Zhifeng Chen. All rights reserved.
import UIKit
class Shape {
    //名称
    var name : String?
    //边数
    var sides : Int?
    //左上角的位置坐标
    var origin : CGPoint?
    //线条颜色
    var lineColor : UIColor = UIColor.red
    //填充颜色
    var fillColor : UIColor = UIColor.green
    //线条宽度
    var lineWidth : CGFloat = 2
    //构造器函数 init
    init(name : String, sides : Int, origin : CGPoint) {
        self.name = name
        self.sides = sides
        self.origin = origin
    }
    //自定义方法 sayHello
    func sayHello(){
        print("Shape is \(name!),sides \(sides!), and originCord is (\(origin!.x),\(origin!.y)) ")
    }
}
class CzfView : UIView {
    //成员变量(属性)shape,其类型为 Shape
    var shape : Shape?
    //重载 UIView 的 draw 方法
    override func draw(_ rect: CGRect) {
        //判断 shape 变量是否为空值 nil
        guard let s = shape else {
            return
        }
        //不为空,则调用 shape 这个实例的方法
        s.sayHello()
    }
}
```

```swift
class ViewController: UIViewController {
    override func viewDidLoad(){
        super.viewDidLoad()
        // Do any additional setup after loading the view, typically from a nib.
        //此处调用 Shape 类,建立一个对象(实例)myShape
        let myShape = Shape(name:"BaseShape", sides: 0, origin: CGPoint(x: 0, y: 0))
        //此处建立了一个 CzfView 的实例 myView
        let myView = CzfView (frame: CGRect (x: 0, y: 0, width: self.view.frame.size.width, height: self.view.frame.size.height))
        //赋值给 myView 中的成员变量(属性)shape
        myView.shape = myShape
        //显示 myView
        self.view.addSubview(myView)
    }
    override func didReceiveMemoryWarning() {
        super.didReceiveMemoryWarning()
        // Dispose of any resources that can be recreated.
    }
}
```

下面通过给 Shape 类增加一个方法 drawBezierPath(),实现画一个矩形的功能。

具体来讲,在 Shape 类中增加一个方法 drawBezierPath(),其代码如下:

```swift
//自定义方法 drawBezierPath 用于画图
func drawBezierPath(){
    //调用贝塞尔曲线函数 UIBezierPath()
    let path = UIBezierPath()
    //圆弧的中心点 center,其坐标为(100,100)
    let center : CGPoint = CGPoint(x: 100, y: 100)
    //圆弧的半径长度 radius,其值为 80
    let radius : CGFloat = 80
    //生成一个圆
    path.addArc(withCenter: center, radius: radius, startAngle: 0, endAngle: CGFloat.pi * 2, clockwise: true)
    //线条宽度为 5
    path.lineWidth = 5
    //线条颜色为 red 红色
    UIColor.red.setStroke()
    //画出这个圆
    path.stroke()
}
```

在 CzfView 这个类中,修改 draw 方法,增加一行:

```swift
s.drawBezierPath()
```

最后在 ViewController 的 ViewDidLoad( )中,清除背景色,不然会出现黑色背景,具体增加的代码如下:

```swift
myView.backgroundColor = UIColor.clear //清除背景色
```

程序修改后的运行效果如图 5-18 所示。

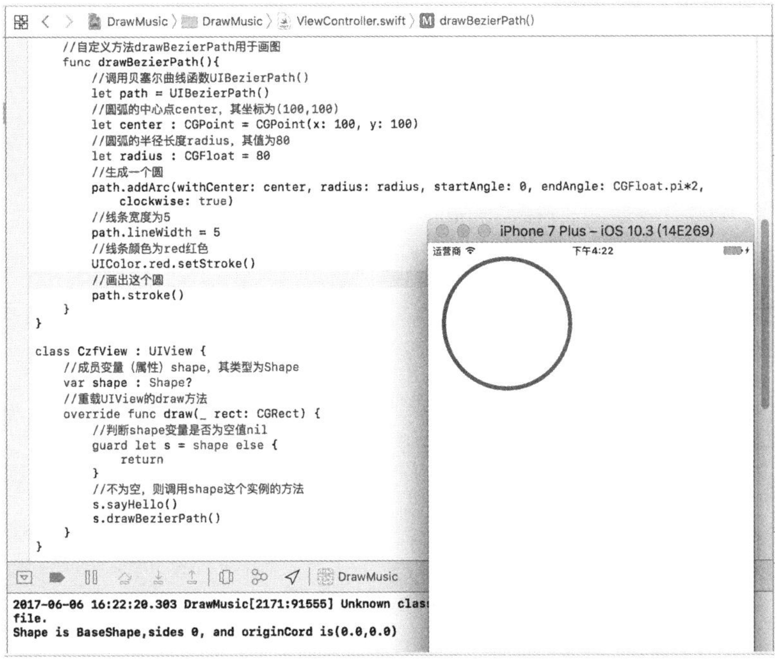

图 5-18 圆的显示

5.4 绘制平面几何图形

平面几何图形:有些几何图形(如线段、角、三角形、长方形、圆等)的各个部分都在同一平面内,它们是平面图形。例如,直线、射线、角、三角形、平行四边形、长方形(正方形)、梯形和圆都是几何图形,这些图形所表示的各个部分都在同一平面内,称为平面图形。平面图形的大小,叫作它们的面积。点形成线,线形成面,面形成体,如图 5-19 所示。例如,有一组对边平行的四边形一定是平面图形。

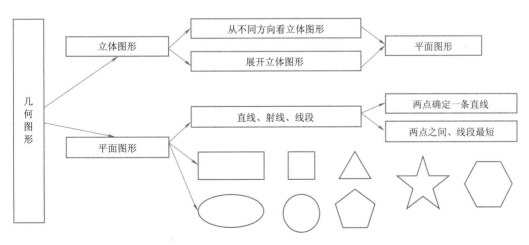

图 5-19 几何图形

在此主要采用 UIKit 绘图方式来实现各种绘图操作,主要是通过 UIBezierPath 这个类来实现的,创建基本的平面几何图形,如线段、正方形、圆、圆弧、长方形、多边形和贝塞尔曲线等,也可以根据自己的设计和计算,画出一些特别的平面图形,如五角星等。

5.4.1 常见图形绘制

UIKit 绘图方式实际是对 Core Graphics 方式的一种简化封装,可以采用面向对象的方式很方便地做各种绘图操作,主要是通过 UIBezierPath 这个类来实现的,创建基于矢量的路径,例如各种直线、曲线、圆等。

1. 基类 Shape

对于所有的图形来讲,可以设计一个基本类,其名称就是 Shape,主要建立所有图形类所共有的基本的属性和方法,当然也包括构造器。

程序代码如下:

```swift
class Shape {
    //名称
    var name : String?
    //左上角的位置坐标
    var origin : CGPoint?
    //线条颜色
    var lineColor : UIColor? = UIColor.red
    //填充颜色
    var fillColor : UIColor? = UIColor.green
    //线条宽度
    var lineWidth : CGFloat? = 5
    //构造器函数 init
    init(name : String, origin : CGPoint) {
        self.name = name
        self.origin = origin
    }
    //便利构造器函数 init
    convenience init(origin : CGPoint) {
        self.init(name: "Shape Bassclass", origin: origin)
    }
    //自定义方法 drawBezierPath 用于画图
    func drawBezierPath(){
        //向控制台输出信息
        print("Draw \(name!)")
    }
}
```

指定构造器(Designated Initializers),对所有没有默认值的非可选属性进行初始化。

便利构造器(Convenience Initializers),是在 init 前加一个关键字 convenience,它为一些属性提供默认值。这样,在初始化时就无须给所有属性赋值。通常要调用类自身的便利构造器或者指定构造器,不管是哪种,最终都要调用指定构造器。

2. 线段类 Line

由平面几何图形基本知识可知,线段是通过连接两个点来形成的。线段类 Line 派生字 Shape 类。Swift 已经提供了基本的画线函数 path.move(to: CGPoint) 和 path.addLine(to: CGPoint),一个是用于移动,一个是从当前点开始画,到结束点为止。

程序代码如下:

```
class Line : Shape {
    //线段的起点
    var start : CGPoint?
    //线段的终点
    var end : CGPoint?
    //构造器
    init(name: String, origin: CGPoint , start : CGPoint , end : CGPoint) {
        super.init(name: name, origin: origin)
        self.start = start
        self.end = end
    }
    convenience init(start : CGPoint , end : CGPoint) {
        self.init(name: "Line", origin: start, start: start, end: end)
    }
    //重载 drawBezierPath 方法
    override func drawBezierPath(){
        //向控制台输出信息
        print("Draw \(name!)")
        //建立一个 UIBezierPath 实例对象
        let path = UIBezierPath()
        //调用实例 path 的 move 方法移动
        path.move(to: start!)
        //调用实例 path 的 addLine 方法画线
        path.addLine(to: end!)
        //设置实例 path 的线条宽度
        path.lineWidth = lineWidth!
        //设置实例 path 的线条终端样式. round 或者.square
        path.lineCapStyle = .round
        //设置实例 path 的线条颜色
        lineColor?.setStroke()
        //画出线条
        path.stroke()
    }
}
```

3. 矩形类 Rectangle

矩形一般可以通过确定左上角的坐标位置以及矩形的宽和高,就可以正确地画出来,正方形就是特殊的矩形,其宽和高相等。或者,知道这个矩形的中心坐标以及矩形的宽和高,通过计算也可以很方便地画出这个矩形。

下面采用 UIKit 中的 UIBezierPath(rect: CGRect)方法来实现矩形的绘制,也可以采用前面画线段的方法来绘制矩形。

程序代码如下：

```swift
class Rectangle : Shape {
    //左上角 采用基类 Shape 中的属性 origin
    //宽度 和 高度,一般可以采用 CGSize
    var size : CGSize?
    //构造器
    init(name: String, origin: CGPoint , size : CGSize) {
        super.init(name: name, origin: origin)
        self.size = size
    }
    convenience init(origin: CGPoint , size : CGSize) {
        self.init(name: "Rectangle", origin: origin, size: size)
    }
    //重载 drawBezierPath 方法
    override func drawBezierPath(){
        //向控制台输出信息
        print("Draw \(name!)")
        //建立一个 UIBezierPath 实例对象
        let path = UIBezierPath(rect: CGRect(origin: origin!, size: size!))
        //设置实例 path 的线条宽度
        path.lineWidth = lineWidth!
        //设置实例 path 的线条颜色
        lineColor?.setStroke()
        //画出线条
        path.stroke()
    }
}
```

在实际编程中，还需要画具有圆角的矩形，这就需要采用 UIKit 中的 UIBezierPath（roundedRect：CGRect, cornerRadius：CGFloat）方法来实现矩形的绘制，这样，需要修改这个矩形类。

程序代码如下：

```swift
class Rectangle : Shape {//左上角 采用基类 Shape 中的属性 origin
    //宽度 和 高度,一般可以采用 CGSize
    var size : CGSize?
    //圆角的大小
    var corner : CGFloat?
    //构造器
    init(name: String, origin: CGPoint , size : CGSize , corner : CGFloat) {
        super.init(name: name, origin: origin)
        self.size = size
        self.corner = corner
    }
    convenience init(origin: CGPoint , size : CGSize, corner : CGFloat = 0) {
        self.init(name: "Rectangle", origin: origin, size: size, corner : corner)
    }
    //重载 drawBezierPath 方法
```

```
override func drawBezierPath(){
//向控制台输出信息
print("Draw \(name!)")
//建立一个 UIBezierPath 实例对象
let path = UIBezierPath(roundedRect: CGRect(origin: origin!, size: size!), cornerRadius: corner!)

//设置实例 path 的线条宽度
path.lineWidth = lineWidth!
//设置实例 path 的线条颜色
lineColor?.setStroke()
//画出线条
path.stroke()
}
}
```

这样在调用的时候用如下代码：

```
let myRect = Rectangle(origin: start, size: CGSize(width: 150, height: 100),corner: 8.0)
```

则会生成一个带圆角的矩形。如果调用方法如下：

```
let myRect = Rectangle(origin: start, size: CGSize(width: 150, height: 100))
```

则会生成一个不带圆角的普通矩形，因为在类的便利构造器中，corner 默认数是 0，也就是相当于圆角为 0。

4. 圆形类 Circle

圆形一般可以通过确定圆所在的中心坐标位置及半径，就可以正确地画出来，圆形也可以看作是宽和高相等的椭圆。或者，知道这个椭圆所外接的矩形，也可以很方便地画出这个矩形。

这里采用 UIKit 中的 UIBezierPath（ovalIn：CGRect）方法来实现椭圆或者圆的绘制。

具体是绘制正圆还是椭圆可以直接通过 width 与 height 来控制，两者相等绘制出来就是正圆，否则就是椭圆。

程序代码如下：

```
class Circle : Shape {
    //圆心坐标
    var center : CGPoint?
    //半径长度
    var raduis : CGFloat?
    //椭圆的宽度 和 高度,一般可以采用 CGSize
    var size : CGSize?
    //构造器
    init(name: String, origin: CGPoint , center : CGPoint , raduis : CGFloat , size : CGSize) {
        super.init(name: name, origin: origin)
        self.center = center
        self.raduis = raduis
        self.size = size
    }
```

```swift
        convenience init(center : CGPoint,raduis : CGFloat) {
            let x = center.x - raduis
            let y = center.y - raduis
            self.init(name: "Circle", origin: CGPoint(x:x,y:y), center: center, raduis: raduis, size: CGSize(width: raduis, height: raduis))
        }
        convenience init(center : CGPoint, size : CGSize) {
            let x = center.x - size.width/2
            let y = center.y - size.height/2
            self.init(name: "Oval/Ellipse", origin: CGPoint(x:x,y:y), center: center, raduis: 0, size: size)
        }
        //重载 drawBezierPath 方法
        override func drawBezierPath(){
            //向控制台输出信息
            print("Draw \(name!)")
            //建立一个 UIBezierPath 实例对象
            let path = UIBezierPath(ovalIn: CGRect(origin: origin!, size: size!))
            //设置实例 path 的线条宽度
            path.lineWidth = lineWidth!
            //设置实例 path 的线条颜色
            lineColor?.setStroke()
            //画出线条
            path.stroke()
        }
    }
```

5. 多边形类 Polygons

多边形的绘制主要依赖 move(to：CGPoint)与 addLine(to：CGPoint)这两个方法,通过不同的组合画出不同的图形。多个点链接起来就是建立一个多边形,如果是三个点,则组成一个三角形。

程序代码如下:

```swift
class Polygons : Shape {
    //多边形主要是通过多个顶点相互链接来绘图
    //顶点数组
    var points : Array<CGPoint>?
    init(name: String, origin: CGPoint , points : Array<CGPoint> ) {
        super.init(name: name, origin: origin)
        self.points = points
    }
    convenience init(points : Array<CGPoint> ) {
        if points.count == 3 {
            self.init(name: "Triangel", origin: points.first!, points: points)
        }
        else if points.count >= 3 {
            self.init(name: "Polygons", origin: points.first!, points: points)
        }
        else {
```

```
            let origin = CGPoint(x: 0, y: 0)
            self.init(name: "Error", origin: origin, points: points)
        }
    }
    //重载 drawBezierPath 方法
    override func drawBezierPath(){
        //向控制台输出信息
        print("Draw \(name!)")
        //如果不能识别,直接返回
        if(name = = "Error") {
            return
        }
        //建立一个 UIBezierPath 实例对象
        let path = UIBezierPath()
        //调用实例 path 的 move 方法移动
        path.move(to: origin!)
        for each in points! {
            //调用实例 path 的 addLine 方法画线
            path.addLine(to: each)
        }
        path.close()
        //设置实例 path 的线条宽度
        path.lineWidth = lineWidth!
        //设置实例 path 的线条颜色
        lineColor?.setStroke()
        //画出线条
        path.stroke()
        path.fill()
    }
}
```

三角形一般常见的是根据 3 个顶点来画出三条线段,组合成一个三角形。三角形和夹角的关系如图 5-20 所示。特殊一点的三角形有等边三角形和直角三角形。

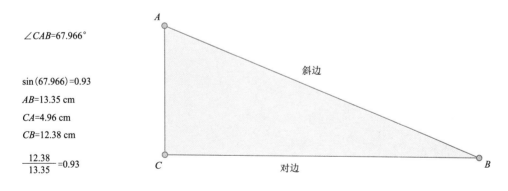

图 5-20　三角形与夹角的关系

5.4.2　五角星的绘制

在做手工时,许多时候大家都用正五角星来装点自己的艺术作品,下面学习一下如何用尺规作图法画正五角星。

在白纸上，以任意一点为圆心，以任意长为半经画圆 O。在圆中画两条互相垂直的圆的直经 AB 和 CD。取线段 OB 的中点 E，连接 CE。以点 E 为圆心，以 CE 长为半经画圆弧，交线段 OA 于点 F。连接 CF，以点 C 为圆心，以 CF 长为半经在圆 O 上依次截取相等的圆弧。连接 CM、CH、GN、GM、NH，就得到正五角星，如图 5-21 所示。

在软件编码时，经常采用量角器法类画五角星：

（1）确定圆心。

（2）确定圆心后，以合适的长度为半径画圆。

（3）过圆心画水平直线和竖直直线。

（4）以竖直直线与圆的上端交点为起点，用量角器顺时针（也可以逆时针）在圆上每隔 72°标出点。

（5）从顶点开始（任一点都可以）隔点连接（顺序连接，得到的是正五边形）。

绘制结果如图 5-22 所示。

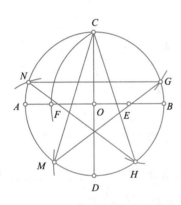

图 5-21　尺规作图法画正五角星

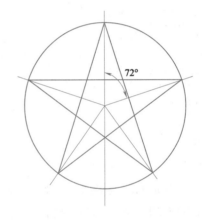

图 5-22　采用量角器，每 72°等分这个圆为五份

程序代码如下：

```swift
class FiveStar : Shape {
    //五角星和正五边形较为类似
    //中心坐标
    var center : CGPoint?
    //半径
    var radius : CGFloat?
    //旋转的角度
    var angel : CGFloat?
    //构造器
    init(name: String, origin: CGPoint , center : CGPoint, radius : CGFloat,angel : CGFloat ) {
        super. init(name: name, origin: origin)
        self. center = center
        self. radius = radius
        self. angel  = angel
    }
```

```
    convenience init(center : CGPoint, radius : CGFloat,angel : CGFloat = 0){
        let x = center.x - radius
        let y = center.y - radius
        self.init(name: "FiveStar", origin: CGPoint(x:x,y:y), center: center, radius: radius, angel: angel)
    }
    //自定义方法 drawBezierPath 用于画五角星
    override func drawBezierPath(){
        //调用贝塞尔曲线函数 UIBezierPath()
        let path = UIBezierPath()
        //五角星旋转顶点
        let i = 360/angel!
        let xzAngle = CGFloat.pi* 2/i
        let xzX = (center?.x)! - sin(xzAngle)* radius!
        let xzY = (center?.y)! - cos(xzAngle)* radius!
        let p1 = CGPoint(x: xzX, y: xzY)
        path.move(to: p1)
        let angle = CGFloat.pi* 4/5
        for i in 1...5 {
            let x = (center?.x)! - sin(CGFloat(i)* angle +xzAngle)* radius!
            let y = (center?.y)! - cos(CGFloat(i)* angle +xzAngle)* radius!
            path.addLine(to: CGPoint(x: x, y: y))
        }
        path.close()
        //线条宽度
        path.lineWidth = lineWidth!
        //线条颜色为 red 红色
        lineColor?.setStroke()
        //画出这个圆
        path.stroke()
    }
}
```

5.5 奏响乐队凯歌

5.5.1 声音播放功能

为了让所有的图形实现声音播放功能,同时每个图形的播放声音文件不同,但其播放的代码一般没有不同,因此,这里重新修改了 Shape 类,为其增加了 2 个属性(var soundPlayer:AVAudioPlayer? 和 var soundFile：String?)和 1 个方法 func playAudio(),同时对构造器也进行了调整。

在 Shape 类中,如果构造器调用的时候没有 soundFile 参数输入,则默认会使用 soundFile：String = "DO.m4a"。

程序代码如下:

```swift
import AVFoundation
class Shape {
    //名称
    var name : String?
    //声音播放器
    var soundPlayer : AVAudioPlayer?
    //声音文件名称
    var soundFile : String?
    //左上角的位置坐标
    var origin : CGPoint?
    //线条颜色
    var lineColor : UIColor? = UIColor.red
    //填充颜色
    var fillColor : UIColor? = UIColor.green
    //线条宽度
    var lineWidth : CGFloat? = 5
    //构造器函数 init
    init(name : String, origin : CGPoint,soundFile : String = "DO.m4a" ) {
        self.name = name
        self.origin = origin
        self.soundFile = soundFile
    }
    //遍历构造器函数 init
    convenience init(origin : CGPoint) {
        self.init(name: "Shape Bassclass", origin: origin)
    }
    //自定义方法 drawBezierPath 用于画图
    func drawBezierPath(){
        //向控制台输出信息
        print("Draw \(name!)")
    }
    func playAudio(){
        print("Play sound:\(soundFile!)")
        let path = Bundle.main.path(forResource: soundFile, ofType: nil)
        let url = URL(fileURLWithPath: path!)
        soundPlayer = try? AVAudioPlayer(contentsOf: url)
        soundPlayer?.play()
    }
}
```

如何使用这个声音播放功能？除了在 Shape 这个类中添加相应的代码外，还需要在 CzfView 中增加调用 s.playAudio()，这样在整个程序启动的时候，就可以听到音符的声音。

程序代码如下：

```swift
// ViewController.swift
// GeometryBand
// Created by Zhifeng Chen on 2017/6/7.
// Copyright ? 2017 年 Zhifeng Chen. All rights reserved.
```

```swift
import UIKit
import Foundation
import AVFoundation
class Shape {
    //名称
    var name : String?
    //声音播放器
    var soundPlayer : AVAudioPlayer?
    //声音文件名称
    var soundFile : String?
    //左上角的位置坐标
    var origin : CGPoint?
    //线条颜色
    var lineColor : UIColor? = UIColor.red
    //填充颜色
    var fillColor : UIColor? = UIColor.green
    //线条宽度
    var lineWidth : CGFloat? = 5
    //构造器函数 init
    init(name : String, origin : CGPoint,soundFile : String = "DO.m4a" ) {
        self.name = name
        self.origin = origin
        self.soundFile = soundFile
    }
    //便利构造器函数 init
    convenience init(origin : CGPoint) {
        self.init(name: "Shape Bassclass", origin: origin)
    }
    //自定义方法 drawBezierPath 用于画图
    func drawBezierPath(){
        //向控制台输出信息
        print("Draw \(name!)")
    }
    func playAudio(){
        print("Play sound:\(soundFile!)")
        let path = Bundle.main.path(forResource: soundFile, ofType: nil)
        let url = URL(fileURLWithPath: path!)
        soundPlayer = try? AVAudioPlayer(contentsOf: url)
        soundPlayer?.play()
    }
}
class Line : Shape {
    //线段的起点
    var start : CGPoint?
    //线段的终点
    var end : CGPoint?
    //构造器
    init(name: String, origin: CGPoint , start : CGPoint , end : CGPoint){
        super.init(name: name, origin: origin)
        self.start = start
```

```
            self.end = end
        }
        convenience init(start : CGPoint , end : CGPoint) {
            self.init(name: "Line", origin: start, start: start, end: end)
        }
        //重载 drawBezierPath 方法
        override func drawBezierPath(){
            //向控制台输出信息
            print("Draw \(name!)")
            //建立一个 UIBezierPath 实例对象
            let path = UIBezierPath()
            //调用实例 path 的 move 方法移动
            path.move(to: start!)
            //调用实例 path 的 addLine 方法画线
            path.addLine(to: end!)
            //设置实例 path 的线条宽度
            path.lineWidth = lineWidth!
            //设置实例 path 的线条终端样式.round 或者.square
            path.lineCapStyle = .round
            //设置实例 path 的线条颜色
            lineColor?.setStroke()
            //画出线条
            path.stroke()
        }
    }
    class Rectangle : Shape {
        //左上角采用基类 Shape 中的属性 origin
        //宽度和高度,一般可以采用 CGSize
        var size : CGSize?
        //圆角的大小
        var corner : CGFloat?
        //构造器
        init(name: String, origin: CGPoint , size : CGSize , corner : CGFloat){
            super.init(name: name, origin: origin)
            self.size = size
            self.corner = corner
        }
        convenience init(origin: CGPoint , size : CGSize, corner : CGFloat = 0){
            self.init(name: "Rectangle", origin: origin, size: size, corner : corner)
        }
        //重载 drawBezierPath 方法画矩形或者正方形
        override func drawBezierPath(){
            //向控制台输出信息
            print("Draw \(name!)")
            //建立一个 UIBezierPath 实例对象
            let path = UIBezierPath(roundedRect: CGRect(origin: origin!, size: size!), cornerRadius: corner!)
            //设置实例 path 的线条宽度
            path.lineWidth = lineWidth!
            //设置实例 path 的线条颜色
```

```swift
        lineColor?.setStroke()
        //画出线条
        path.stroke()
    }
}
class Circle : Shape{
    //圆心坐标
    var center : CGPoint?
    //半径长度
    var raduis : CGFloat?
    //椭圆的宽度和高度,一般可以采用CGSize
    var size : CGSize?
    //构造器
    init(name: String, origin: CGPoint , center : CGPoint , raduis : CGFloat, size : CGSize){
        super.init(name: name, origin: origin)
        self.center = center
        self.raduis = raduis
        self.size = size
    }
    convenience init(center : CGPoint,raduis : CGFloat){
        let x = center.x - raduis
        let y = center.y - raduis
        self.init(name: "Circle", origin: CGPoint(x:x,y:y), center: center, raduis: raduis, size: CGSize(width: raduis, height: raduis))
    }
    convenience init(center : CGPoint, size : CGSize){
        let x = center.x - size.width/2
        let y = center.y - size.height/2
        self.init(name: "Oval/Ellipse", origin: CGPoint(x:x,y:y), center: center, raduis: 0, size: size)
    }
    //重载drawBezierPath方法画圆或者椭圆
    override func drawBezierPath(){
        //向控制台输出信息
        print("Draw \(name!)")
        //建立一个UIBezierPath实例对象
        let path = UIBezierPath(ovalIn: CGRect(origin: origin!, size: size!))
        //设置实例path的线条宽度
        path.lineWidth = lineWidth!
        //设置实例path的线条颜色
        lineColor?.setStroke()
        //画出线条
        path.stroke()
    }
}
class Polygons : Shape {
    //多边形主要是通过多个个顶点相互链接来绘图
    //顶点数组
    var points : Array<CGPoint>?
    //构造器
```

```swift
        init(name: String, origin: CGPoint , points : Array<CGPoint> ){
            super.init(name: name, origin: origin)
            self.points = points
        }
    convenience init(points : Array<CGPoint> ){
        if points.count == 3 {
            self.init(name: "Triangel", origin: points.first!, points: points)
        }
        else if points.count >= 3 {
            self.init(name: "Polygons", origin: points.first!, points: points)
        }
        else {
            let origin = CGPoint(x: 0, y: 0)
            self.init(name: "Error", origin: origin, points: points)
        }
    }
//重载 drawBezierPath 方法画多边形
override func drawBezierPath(){
    //向控制台输出信息
    print("Draw \(name!)")
    //如果不能识别,直接返回
    if(name == "Error") {
        return
    }
    //建立一个 UIBezierPath 实例对象
    let path = UIBezierPath()
    //调用实例 path 的 move 方法移动
    path.move(to: origin!)
    for each in points! {
        //调用实例 path 的 addLine 方法画线
        path.addLine(to: each)
    }
    path.close()
    //设置实例 path 的线条宽度
    path.lineWidth = lineWidth!
    //设置实例 path 的线条颜色
    lineColor?.setStroke()
    //画出线条
    path.stroke()
    path.fill()
    }
}
class FiveStar : Shape {
    //五角星和正五边形较为类似
    //中心坐标
    var center : CGPoint?
    //半径
    var radius : CGFloat?
    //旋转的角度
    var angel : CGFloat?
```

```swift
        //构造器
        init(name: String, origin: CGPoint , center : CGPoint, radius : CGFloat,angel : CGFloat ) {
            super. init(name: name, origin: origin)
            self. center = center
            self. radius = radius
            self. angel = angel
        }
        convenience init(center : CGPoint, radius : CGFloat,angel : CGFloat = 0) {
            let x = center. x - radius
            let y = center. y - radius
            self. init(name: "FiveStar", origin: CGPoint (x:x, y:y), center: center, radius: radius, angel: angel)
        }
        //自定义方法 drawBezierPath 用于画五角星
        override func drawBezierPath(){
            //调用贝塞尔曲线函数 UIBezierPath()
            let path = UIBezierPath()
            //五角星旋转顶点
            let i = 360/angel!
            let xzAngle = CGFloat. pi* 2/i
            let xzX = (center? . x)! - sin(xzAngle)* radius!
            let xzY = (center? . y)! - cos(xzAngle)* radius!
            let p1 = CGPoint (x: xzX, y: xzY)
            path. move (to: p1)
            let angle = CGFloat. pi* 4/5
            for i in 1...5 {
                let x = (center? . x)! - sin(CGFloat(i)* angle + xzAngle)* radius!
                let y = (center? . y)! - cos(CGFloat(i)* angle + xzAngle)* radius!
                path. addLine (to: CGPoint (x: x, y: y))
            }
            path. close ()
            //线条宽度
            path. lineWidth = lineWidth!
            //线条颜色为 red 红色
            lineColor? . setStroke ()
            //画出这个圆
            path. stroke ()
        }
}
class CzfView : UIView {
    //成员变量(属性) shape,其类型为 Shape
    var shape : Shape?
    //重载 UIView 的 draw 方法
    override func draw(_ rect: CGRect) {
        //判断 shape 变量是否为空值 nil
        guard let s = shape else {
            return
        }
        //不为空,则调用 shape 这个实例的方法
        s. drawBezierPath()
```

```swift
            s.playAudio()
        }
    }
class ViewController: UIViewController {
    override func viewDidLoad(){
        super.viewDidLoad()
        // Do any additional setup after loading the view, typically from a nib.
        //此处调用 FiveStar 类,建立一个对象(实例)star
        //五角星的中心坐标为(180,180)
        let center = CGPoint(x: 180, y: 180)
        //五角星的半径设定了90,旋转角度为15度
        let star = FiveStar(center: center, radius: 90, angel: 15)
        //此处建立了一个 CzfView 的实例 myView
        let myView = CzfView(frame: CGRect(x: 0, y: 0, width: self.view.frame.size.width, height: self.view.frame.size.height))
        //清除背景色
        myView.backgroundColor = UIColor.clear
        //赋值给 myView 中的成员变量(属性)shape
        myView.shape = star
        //显示 myView
        self.view.addSubview(myView)
    }
    override func didReceiveMemoryWarning(){
        super.didReceiveMemoryWarning()
        // Dispose of any resources that can be recreated.
    }
}
```

5.5.2 多个图形显示

如何才能实现多个图形同时显示呢？这就需要在 CzfView 中增加属性,也就是一个能保存多个图形实例的数组,同时对 draw()方法进行修改。同时需要修改 ViewController.swift 中的 ViewDidLoad()方法中的代码。

程序代码如下：

```swift
class CzfView : UIView {
    //成员变量(属性)shapes,其类型为 Array<Shape>
    private var shapes : Array<Shape> = []
    //重载 UIView 的 draw 方法
    override func draw(_ rect: CGRect) {
        //调用 shapes 这个数组中的每个实例的方法
        for s in shapes {
            s.drawBezierPath()
            s.playAudio()
        }
    }
    //增加实例到数组 shapes 中
    func add(shape : Shape) {
```

```
            shapes.append(shape)
        }
}
class ViewController: UIViewController {
    override func viewDidLoad() {
        super.viewDidLoad()
        // Do any additional setup after loading the view, typically from a nib.
        //此处调用 FiveStar 类,建立一个对象(实例)star
        //五角星的中心坐标为(180,180)
        let starCenter = CGPoint(x: 180, y: 180)
        //五角星的半径设定了 90,旋转角度为 15°
        let star = FiveStar(center: starCenter, radius: 50, angel: 45)
        star.lineColor = UIColor.blue
        //此处调用 Circle 类,建立一个对象 oval
        let ovalCenter = CGPoint(x: 100, y: 300)
        let ovalSize = CGSize(width: 100, height: 60)
        let oval = Circle(center: ovalCenter, size: ovalSize)
        //此处调用 Rectangle 类,建立一个对象 rect
        let rectOrigin = CGPoint(x: 60, y: 50)
        let rectSize = CGSize(width: 100, height: 50)
        let rect = Rectangle(origin: rectOrigin, size: rectSize, corner: 6)
        rect.lineColor = UIColor.gray
        //此处建立了一个 CzfView 的实例 myView
        let myView = CzfView (frame: CGRect (x: 0, y: 0, width: self.view.frame.size.width, height: self.view.frame.size.height))
        //清除背景色
        myView.backgroundColor = UIColor.clear
        //赋值给 myView 中的成员变量(属性)shapes
        myView.add(shape: star)
        myView.add(shape: oval)
        myView.add(shape: rect)
        //显示 myView
        self.view.addSubview(myView)
    }
}
```

现在图形显示的同时会发出一个相同的声音,那么如何让不同的类的实例发出不同的声音呢？这就可以通过修改各个类的构造器来实现。

例如,FiveStar 类的构造器需要修改如下：

```
//构造器
    init(name: String, origin: CGPoint , center : CGPoint, radius : CGFloat, angel : CGFloat, soundFile : String = "SO.m4a") {
        super.init(name: name, origin: origin, soundFile: soundFile)
        self.center = center
        self.radius = radius
        self.angel = angel
    }
```

也就是说，让每个类的构造器增加一个参数 soundFile，这个参数可以有一个默认值 soundFile：String = "SO.m4a"，这样保证每个类都有一个不同的声音。

5.5.3 触摸事件与虚线

在 CzfView 中重载 touchesBegan()方法，根据获得触摸位置的坐标，通过在所有的 Shape 或者其派生的类的实例进行判断，当前触摸的位置是否位于 path 范围。也就是说，当 path 是一个矩形的时候，判断是否在这个矩形内；如果 path 是一个圆形的时候，判断是否在这个圆内。具体在 Shape 类中增加一个属性 selectedFlag，和一个自定义方法 func isSelected (point：CGPoint) -> Bool。

程序代码如下：

```swift
// ViewController.swift
// GeometryBand
// Created by Zhifeng Chen on 2017/6/7.
// Copyright ? 2017 年 Zhifeng Chen. All rights reserved.
import UIKit
import Foundation
import AVFoundation

class Shape {
    //名称
    var name : String?
    //声音播放器
    var soundPlayer : AVAudioPlayer?
    //声音文件名称
    var soundFile : String?
    //UIBezierPath
    var path : UIBezierPath?
    //selected?
    var selectedFlag : Bool = false
    //左上角的位置坐标
    var origin : CGPoint?
    //线条颜色
    var lineColor : UIColor? = UIColor.red
    //填充颜色
    var fillColor : UIColor? = UIColor.green
    //线条宽度
    var lineWidth : CGFloat? = 5
    //构造器函数 init
    init(name : String, origin : CGPoint,soundFile : String = "DO.m4a" ) {
        self.name = name
        self.origin = origin
        self.soundFile = soundFile
    }
    //便利构造器函数 init
    convenience init(origin : CGPoint) {
        self.init(name: "Shape Bassclass", origin: origin)
    }
```

```swift
        //自定义方法 drawBezierPath 用于画图
        func drawBezierPath(){
            //向控制台输出信息
            print("Draw \(name!)")
        }
        func playAudio(){
            print("Play sound:\(soundFile!)")
            let path = Bundle.main.path(forResource: soundFile, ofType: nil)
            let url = URL(fileURLWithPath: path!)
            soundPlayer = try? AVAudioPlayer(contentsOf: url)
            soundPlayer?.play()
        }
        func isSelected(point : CGPoint) -> Bool{
            if (path?.contains(point))!{
                selectedFlag = true
                return true
            }
            else {
                selectedFlag = false
                return false
            }
        }
    }
    class Line : Shape {
        //线段的起点
        var start : CGPoint?
        //线段的终点
        var end : CGPoint?
        //构造器
        init(name: String, origin: CGPoint , start : CGPoint , end : CGPoint, soundFile : String = "FA.m4a") {
            super.init(name: name, origin: origin,soundFile:soundFile)
            self.start = start
            self.end = end
        }
        convenience init(start : CGPoint , end : CGPoint) {
            self.init(name: "Line", origin: start, start: start, end: end)
        }
        //重载 drawBezierPath 方法
        override func drawBezierPath(){
            //向控制台输出信息
            print("Draw \(name!)")
            //建立一个 UIBezierPath 实例对象
            path = UIBezierPath()
            //调用实例 path 的 move 方法移动
            path?.move(to: start!)
            //调用实例 path 的 addLine 方法画线
            path?.addLine(to: end!)
            //设置实例 path 的线条宽度
            path?.lineWidth = lineWidth!
```

```swift
        //设置实例 path 的线条终端样式.round 或者.square
        path?.lineCapStyle = .round
        //设置实例 path 的线条颜色
        lineColor?.setStroke()
        //画出线条
        path?.stroke()
    }
}
class Rectangle : Shape {
    //左上角采用基类 Shape 中的属性 origin
    //宽度和高度,一般可以采用 CGSize
    var size : CGSize?
    //圆角的大小
    var corner : CGFloat?
    //构造器
    init(name: String, origin: CGPoint , size : CGSize , corner : CGFloat,soundFile : String = "LA.m4a") {
        super.init(name: name, origin: origin,soundFile:soundFile)
        self.size = size
        self.corner = corner
    }
    convenience init(origin: CGPoint , size : CGSize, corner : CGFloat = 0) {
        self.init(name: "Rectangle", origin: origin, size: size, corner : corner)
    }
    //重载 drawBezierPath 方法画矩形或者正方形
    override func drawBezierPath(){
        //向控制台输出信息
        print("Draw \(name!)")
        //建立一个 UIBezierPath 实例对象
        path = UIBezierPath(roundedRect: CGRect(origin: origin!, size: size!), cornerRadius: corner!)
        if selectedFlag {
            let dashes: [CGFloat] = [1,3]
            path?.setLineDash(dashes, count: dashes.count, phase: 0)
        }
        //设置实例 path 的线条宽度
        path?.lineWidth = lineWidth!
        //设置实例 path 的线条颜色
        lineColor?.setStroke()
        //画出线条
        path?.stroke()
    }
}
class Circle : Shape {
    //圆心坐标
    var center : CGPoint?
    //半径长度
    var raduis : CGFloat?
    //椭圆的宽度和高度,一般可以采用 CGSize
    var size : CGSize?
```

```swift
        //构造器
        init(name: String, origin: CGPoint , center : CGPoint , raduis : CGFloat , size : CGSize ,
soundFile : String = "MI.m4a") {
            super.init(name: name, origin: origin,soundFile:soundFile)
            self.center = center
            self.raduis = raduis
            self.size = size
        }
        convenience init(center : CGPoint,raduis : CGFloat) {
            let x = center.x - raduis
            let y = center.y - raduis
            self.init(name: "Circle", origin: CGPoint(x:x,y:y), center: center, raduis: raduis,
size: CGSize(width: raduis, height: raduis))
        }
        convenience init(center : CGPoint, size : CGSize) {
            let x = center.x - size.width/2
            let y = center.y - size.height/2
            self.init(name: "Oval/Ellipse", origin: CGPoint(x:x,y:y), center: center, raduis: 0,
size: size)
        }
        //重载 drawBezierPath 方法画圆或者椭圆
        override func drawBezierPath(){
            //向控制台输出信息
            print("Draw \(name!)")
            //建立一个 UIBezierPath 实例对象
            path = UIBezierPath(ovalIn: CGRect(origin: origin!, size: size!))
            if selectedFlag {
                let dashes: [CGFloat] = [1,3]
                path?.setLineDash(dashes, count: dashes.count, phase: 0)
            }
            //设置实例 path 的线条宽度
            path?.lineWidth = lineWidth!
            //设置实例 path 的线条颜色
            lineColor?.setStroke()
            //画出线条
            path?.stroke()
        }
}

class Polygons : Shape {
    //多边形主要是通过多个个顶点相互链接来绘图
    //顶点数组
    var points : Array<CGPoint>?
    //构造器
    init(name: String, origin: CGPoint , points : Array<CGPoint> ,soundFile : String = "RE.m4a"){
        super.init(name: name, origin: origin,soundFile: soundFile)
        self.points = points
    }
    convenience init(points : Array<CGPoint> ) {
        if points.count == 3 {
```

```swift
            self.init(name: "Triangel", origin: points.first!, points: points)
        }
        else if points.count >= 3 {
            self.init(name: "Polygons", origin: points.first!, points: points)
        }
        else {
            let origin = CGPoint(x: 0, y: 0)
            self.init(name: "Error", origin: origin, points: points)
        }
    }
    //重载 drawBezierPath 方法画多边形
    override func drawBezierPath(){
        //向控制台输出信息
        print("Draw \(name!)")
        //如果不能识别,直接返回
        if(name == "Error") {
            return
        }
        //建立一个 UIBezierPath 实例对象
        path = UIBezierPath()
        //调用实例 path 的 move 方法移动
        path?.move(to: origin!)
        for each in points! {
            //调用实例 path 的 addLine 方法画线
            path?.addLine(to: each)
        }
        path?.close()
        //设置实例 path 的线条宽度
        path?.lineWidth = lineWidth!
        //设置实例 path 的线条颜色
        lineColor?.setStroke()
        //画出线条
        path?.stroke()
        path?.fill()
    }

}

class FiveStar : Shape {
    //五角星和正五边形较为类似
    //中心坐标
    var center : CGPoint?
    //半径
    var radius : CGFloat?
    //旋转的角度
    var angel : CGFloat?
    //构造器
    init(name: String, origin: CGPoint , center : CGPoint, radius : CGFloat, angel : CGFloat, soundFile : String = "SO.m4a") {
        super.init(name: name, origin: origin, soundFile: soundFile)
```

```swift
            self.center = center
            self.radius = radius
            self.angel = angel
        }
        convenience init(center : CGPoint, radius : CGFloat,angel : CGFloat = 0) {
            let x = center.x - radius
            let y = center.y - radius
            self.init(name: "FiveStar", origin: CGPoint(x:x,y:y), center: center, radius: radius, angel: angel)
        }
        //自定义方法 drawBezierPath 用于画五角星
        override func drawBezierPath(){
            //向控制台输出信息
            print("Draw \(name!)")
            //调用贝塞尔曲线函数 UIBezierPath()
            path = UIBezierPath()
            //五角星旋转顶点
            let i = 360/angel!
            let xzAngle = CGFloat.pi * 2/i
            let xzX = (center?.x)! - sin(xzAngle)* radius!
            let xzY = (center?.y)! - cos(xzAngle)* radius!
            let p1 = CGPoint(x: xzX, y: xzY)
            path?.move(to: p1)
            let angle = CGFloat.pi * 4/5
            for i in 1...5 {
                let x = (center?.x)! - sin(CGFloat(i)* angle + xzAngle)* radius!
                let y = (center?.y)! - cos(CGFloat(i)* angle + xzAngle)* radius!
                path?.addLine(to: CGPoint(x: x, y: y))
            }
            path?.close()
            if selectedFlag {
                let dashes: [CGFloat] = [1,3]
                path?.setLineDash(dashes, count: dashes.count, phase: 0)
            }
            //线条宽度
            path?.lineWidth = lineWidth!
            //线条颜色为 red 红色
            lineColor?.setStroke()
            //画出这个圆
            path?.stroke()
        }
    }

    class CzfView : UIView {
        //成员变量(属性)shapes,其类型为 Array<Shape>
        private var shapes : Array<Shape> = []
        //重载 UIView 的 draw 方法
        override func draw(_ rect: CGRect) {
            //调用 shapes 这个数组中的每个实例的方法
            for s in shapes {
```

```swift
            s.drawBezierPath()
        }
    }
    //增加实例到数组 shapes 中
    func add(shape : Shape) {
        shapes.append(shape)
    }
    //触摸事件
    override func touchesBegan(_ touches: Set<UITouch>, with event: UIEvent?) {
        // 获得 UITouch 集合
        let touch:UITouch = touches.first! as UITouch
        
        // 获得触摸所在位置的坐标
        let point = touch.location(in: self)
        //调用 shapes 这个数组中的每个实例的方法
        for s in shapes {
            //如果被选中
            if s.isSelected(point: point) {
                //声音播放
                s.playAudio()
                //更新屏幕显示
                self.setNeedsDisplay()
            }
        }
    }
}

class ViewController: UIViewController {
    
    override func viewDidLoad() {
        super.viewDidLoad()
        // Do any additional setup after loading the view, typically from a nib.
        //此处调用 FiveStar 类,建立一个对象(实例) star
        //五角星的中心坐标为(180,180)
        let starCenter = CGPoint(x: 180, y: 180)
        //五角星的半径设定了 90,旋转角度为 15 度
        let star = FiveStar(center: starCenter, radius: 50, angel: 45)
        star.lineColor = UIColor.blue
        //此处调用 Circle 类,建立一个对象 oval
        let ovalCenter = CGPoint(x: 100, y: 300)
        let ovalSize = CGSize(width: 100, height: 60)
        let oval = Circle(center: ovalCenter, size: ovalSize)
        //此处调用 Rectangle 类,建立一个对象 rect
        let rectOrigin = CGPoint(x: 60, y: 50)
        let rectSize = CGSize(width: 100, height: 50)
        let rect = Rectangle(origin: rectOrigin, size: rectSize, corner: 6)
        rect.lineColor = UIColor.gray
        //此处建立了一个 CzfView 的实例 myView
        let myView = CzfView (frame: CGRect (x: 0, y: 0, width: self.view.frame.size.width, height: self.view.frame.size.height))
```

```
        //清除背景色
        myView.backgroundColor = UIColor.clear
        //赋值给myView中的成员变量(属性)shape
        myView.add(shape: star)
        myView.add(shape: oval)
        myView.add(shape: rect)
        //显示myView
        self.view.addSubview(myView)
    }
}
```

程序运行效果如图 5-23 所示。

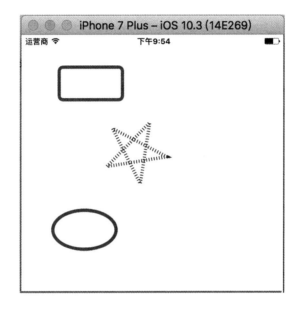

图 5-23　触摸事件与虚线

或许你觉得这个和一个乐队还差很多,没有关系,这只留给你了一片空间,希望你能为乐队扩展成员,组建一个拥有 7 个图形的平面几何图形乐队。

第 6 章 创建机器人聊天室

和传统的聊天室不同,我们是和网络另一头的机器人来聊天,用户可以提问题,可以畅所欲言,因为机器人有的时候比人更善解人意。

6.1 表视图(UITableView)

UITableView 继承自 UIScrollView。一个表视图可以由多个分段(Section)组成,每个分段可以有一个头和尾。很多情况下表视图只有一个分段,而且不显示头尾。表视图本身也可以有一个头(显示在第一个分段之前)和一个尾(显示在最后一个分段之后)。一个表视图的整体元素结构示意如图 6-1 所示。

UITableView 有以下两种样式(UITableViewStyle):Plain(普通样式)和 Grouped(分组样式)。上述两种风格本质无区别,只是显示样式不同而已。

表视图分割线(Separator),有以下 3 种样式(UITableViewCellSeparatorStyle):None(无分割线)、SingleLine(单线条)和 SingleLineEtched(带浮雕效果的线条)。

表视图单元格(Cell),有以下 4 种显示样式(UITableViewCellStyle):

(1) Default:左侧显示 textLabel,不显示 detailTextLabel,最左边可选显示 imageView。

(2) Value1:左侧显示 textLabel,右侧显示 detailTextLabel,最左边可选显示 imageView。

(3) Value2:左侧依次显示 textLabel、detailTextLabel,最左边可选显示 imageView。

(4) Subtitle:左侧上方显示 textLabel,左侧下方显示 detailTextLabel,最左边可选显示 imageView。

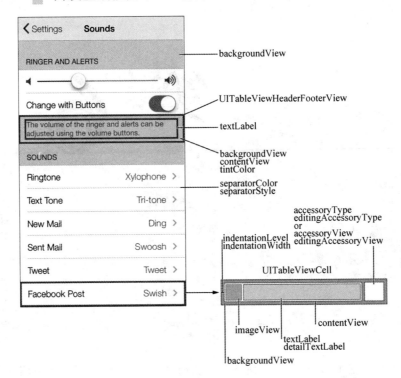

图 6-1 表视图的整体元素结构

Cell 有以下 4 种选中样式（UITableViewCellSelectionStyle）：None、Blue、Gray 和 Default。附属图形（Accessory），有以下 5 种样式（UITableViewCellAccessoryType）：

（1）None：无附属图形。

（2）DisclosureIndicator：小箭头。

（3）DetailDisclosureButton：详细信息按钮 + 指向右侧的箭头。

（4）Checkmark：勾号。

（5）DetailButton：详细信息按钮。

iOS 遵循 MVC 设计模式，很多操作通过代理和外界沟通，UITableView 同理实现了以下两种协议：UITableViewDelegate 和 UITableViewDataSource。

其中，UITableViewDataSource 至少要实现以下两个方法：numberOfRowsInSection（确定表格每个分区拥有多少行）和 cellForRowAtIndexPath（提供一个表格显示用的 Cell）；而 UITableViewDelegate 没有必须实现的方法。

很多时候，一个 UIViewController 中只有一个 UITableView，因此苹果直接提供了一个 UITableViewController，这个控制器实现了 UITableView 数据源和代理协议，内部定义了一个 TableView 属性供外部访问，同时自动铺满整个屏幕、自动伸缩以方便开发。如果需要用到 TableView 时不充满全屏，应该使用 UIViewController 自己创建和维护 UITableView。

6.1.1 表视图的最简单使用

最简单的表视图主要包括：一个存放表格内容的数组 dataArray；一个 UITableView 变量

myTableView；在 ViewController 类声明中继承协议 UITableViewDataSource 和 UITableViewDelegate，并设置委托为 self；数据委托方法 func tableView(_ tableView：UITableView，numberOfRowsInSection section：Int) -> Int 和 func tableView(_ tableView：UITableView，cellForRowAt indexPath：IndexPath) -> UITableViewCell 等。

主要步骤：建立一个 Single View 工程，然后在 ViewController.swift 中输入代码。

程序代码如下：

```swift
//ViewController.swift
//TableTuringChat
//Created by Zhifeng Chen on 2017/6/4.
//Copyright 2017年 Zhifeng Chen. All rights reserved.
import UIKit
class ViewController: UIViewController ,UITableViewDataSource,UITableViewDelegate {
    //variable
    //表格每个单元格显示的内容数组
    var dataArray:Array < String > = ["Chen Zhifeng","Zhang Bo","Chang Yuxiang","Hu Yinting"]
    //建立一个 myTablview 的表格变量
    var myTableView : UITableView!
    override func viewDidLoad() {
        super.viewDidLoad()
        // Do any additional setup after loading the view, typically from a nib.
        //建立一个和屏幕大小一致的矩形框
        let rect = CGRect(x: 0, y: 20, width: self.view.frame.size.width, height: self.view.frame.size.height-20)
        //根据矩形框大小生成一个表格
        myTableView = UITableView(frame: rect)
        //设置表格的代理
        myTableView.delegate = self
        //设置表格的数据代理
        myTableView.dataSource = self
        //显示表格
        self.view.addSubview(myTableView)
    }
    //Mark——tableView Datasource
    //确定表格每个分组有多少行？默认分组为1个
    func tableView(_ tableView: UITableView, numberOfRowsInSection section: Int) -> Int {
        return dataArray.count
    }
    //确定表格每个单元格的样式，和每个单元格显示的内容
    func tableView (_ tableView: UITableView, cellForRowAt indexPath: IndexPath) -> UITableViewCell {
        let cell = UITableViewCell(style: UITableViewCellStyle.default, reuseIdentifier: "Cell")
        let title = dataArray[indexPath.row]
        cell.textLabel?.text = title
        return cell
    }
}
```

程序运行效果如图 6-2 所示。

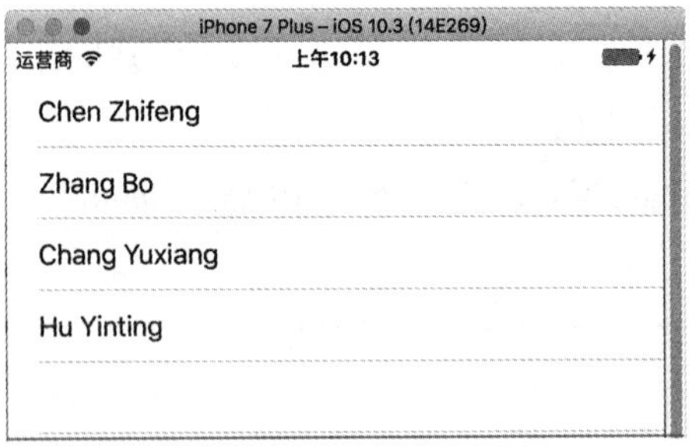

图 6-2　最简单表格效果

下面采用苹果直接提供的 UITableViewController 来建立一个简单的表格，这个控制器实现了 UITableView 数据源和代理协议，内部定义了一个 tableView 属性供外部访问，同时自动铺满整个屏幕、自动伸缩等。

1. 建立一个工程 TableViewSimple

在 macOS 中找到 Xcode，然后运行。在 Xcode 的欢迎界面中选择新建一个 Xcode 工程。从工程模板选择开发 iOS 应用程序，然后选择 Single View Application。在工程参数中主要包括输入工程的名称（Product Name），填写 TableSimple，选择存放在桌面上。

2. 选择 Main. storyboard，拖放 TableViewController 组件

从左边菜单选择 Main. storyboard 文件，单击，在编辑区域会出现手机可视化设计界面。其中只有一个 ViewController 界面，从右边功能区域选择一个 TableViewController 组件，拖放到 Main. storyboard 界面中，如图 6-3 所示。

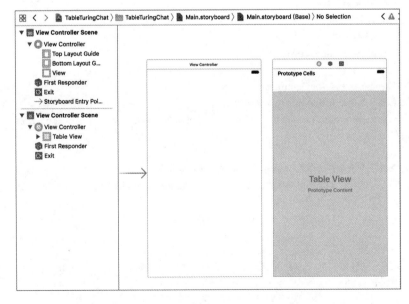

图 6-3　Main. storyboard 中的 ViewController 和 TableViewController

3. 将 TableViewController 设置为启动项

将原来指向 ViewController 的箭头拖动到指向 TableViewController，即可将 TableViewController 设置为启动项，如图 6-4 所示。

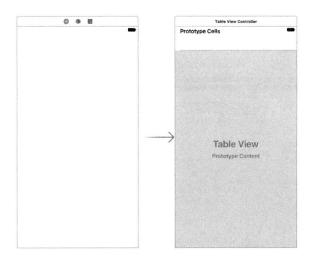

图 6-4　TableViewController 设置为启动项

或者，选择 TableViewController，找到属性查看器，选中"Is Initial View Controller"，如图 6-5 所示。

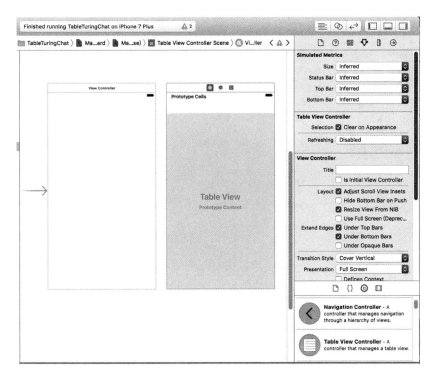

图 6-5　启动项的设置

4. 在工程中新建一个文件

右击左侧工程文件夹，在弹出的快捷菜单中选择 New File 命令，如图 6-6 所示。

图 6-6　工程新建文件

5. 为新文件选择模版

在打开的对话框中选择 Cocoa Touch Class 命令，如图 6-7 所示。

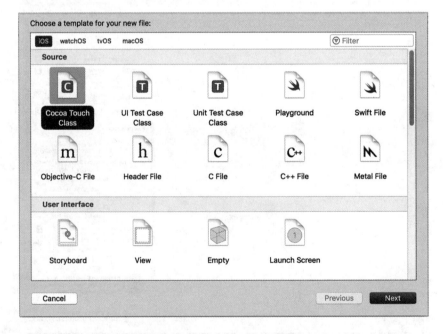

图 6-7　选择 Cocoa Touch Class

6. 为新文件设置选项

在弹出对话框的 Subclass of 下拉列表框中选择 UITableViewController，在 Class 文本框输入名称：MyTableViewController，如图 6-8 所示。

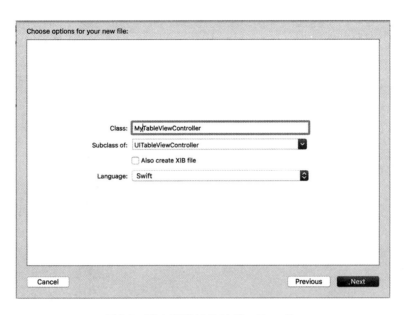

图 6-8　建立新类 MyTableViewController

7. 选择默认位置，退出对话框

在弹出的对话框中，直接单击确定，将退出对话框，返回主界面，显示刚才新建的文件 MyTableViewController.swift 内容，如图 6-9 所示。

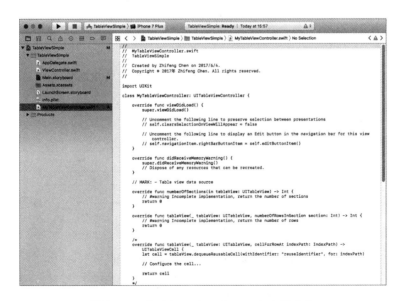

图 6-9　MyTableViewController.swift 文件内容

8. 选择 Main. storyboard，关联类和界面组件

重新选择 Main. storyboard，然后选中 TableViewController，在右侧选择标示查看器，在其中的 Class 中，选择刚才新建文件 MyTableViewController，如图 6-10 所示。

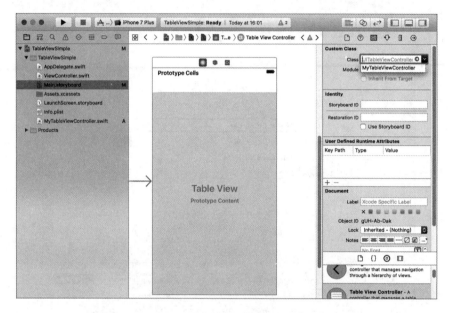

图 6-10　关联 MyTableViewController 类

9. 选择 TableViewCell，设置其标示为 Cell

选择表格的单元格 TableViewCell，在属性查看器中，设置其标示为 Cell，如图 6-11 所示。

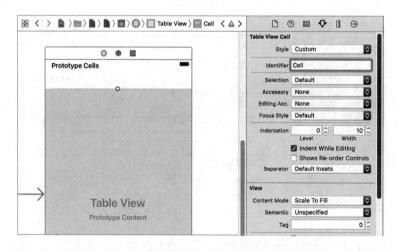

图 6-11　表格单元格名称设置为 Cell

10. 修改 MyTableViewController. swift 代码，运行

修改文件 MyTableViewController. swift 中的代码。

程序代码如下:

```swift
//  MyTableViewController.swift
//  TableViewSimple
//  Created by Zhifeng Chen on 2017/6/4.
//  Copyright 2017年 Zhifeng Chen. All rights reserved.
import UIKit
class MyTableViewController: UITableViewController {
    //variable
    //表格每个单元格显示的内容数组
    var dataArray : Array<String> = ["Chen Zhifeng","Zhang Bo","Chang Yuxiang","Hu Yinting"]
    // MARK: - Table view data source
    override func numberOfSections(in tableView: UITableView) -> Int {
        // #warning Incomplete implementation, return the number of sections
        return 1
    }
    override func tableView(_ tableView: UITableView, numberOfRowsInSection section: Int) -> Int {
        // #warning Incomplete implementation, return the number of rows
        return dataArray.count
    }
    override func tableView(_ tableView: UITableView, cellForRowAt indexPath: IndexPath) -> UITableViewCell {
        let cell = tableView.dequeueReusableCell(withIdentifier: "Cell", for: indexPath)
        // Configure the cell...
        cell.textLabel?.text = dataArray[indexPath.row]
        return cell
    }
}
```

用户可以自行对比一下以上两种方法的代码。另外,如果要使用静态表格,就只能采用 TableViewController。

6.1.2 表视图的一般使用

这里准备采用表格来显示学生信息,主要包括学生姓名 name、特点 character、头像照片 headpic 等。采用了 Dictionary 来保存一个学生数据,多个学生数据保存到一个数组。

程序代码如下:

```swift
//  ViewController.swift
//  TableTuringChat
//  Created by Zhifeng Chen on 2017/6/4.
//  Copyright 2017年 Zhifeng Chen. All rights reserved.
import UIKit
class ViewController: UIViewController ,UITableViewDataSource,UITableViewDelegate {
    //variable
    //表格每个单元格显示的内容数组
    var dataArray : Array<Dictionary<String,String>> = [
        ["name":"Zhang Bo","character":"A good boy with slim body","headpic":"head01.png"],
        ["name":"Chang Yuxiang","character":"A good boy and strong enough","headpic":"head02.png"],
```

```swift
         ["name":"Hu Yinting","character":"A pretty girl and simple temper","headpic":"head03.png"]
    ]
    //建立一个myTablview的表格变量
    var myTableView : UITableView!
    override func viewDidLoad() {
        super.viewDidLoad()
        // Do any additional setup after loading the view, typically from a nib.
        //建立一个和屏幕大小一致的矩形框
        let rect = CGRect(x: 0, y: 20, width: self.view.frame.size.width, height: self.view.frame.size.height-20)
        //根据矩形框大小生成一个表格
        myTableView = UITableView(frame: rect)
        //设置表格的代理
        myTableView.delegate = self
        //设置表格的数据代理
        myTableView.dataSource = self
        //显示表格
        self.view.addSubview(myTableView)
    }
    //Mark——tableView Datasource
    //确定表格每个分组有多少行？默认分组为1个
    func tableView(_ tableView: UITableView, numberOfRowsInSection section: Int) -> Int {
        return dataArray.count
    }
    //确定表格每个单元格的样式,和每个单元格显示的内容
    func tableView(_ tableView: UITableView, cellForRowAt indexPath: IndexPath) -> UITableViewCell {
        let cell = UITableViewCell(style: UITableViewCellStyle.subtitle, reuseIdentifier: "Cell")
        let dic = dataArray[indexPath.row]
        cell.textLabel?.text = dic["name"]
        cell.detailTextLabel?.text = dic["character"]
        cell.imageView?.image = UIImage(named: dic["headpic"]!)
        return cell
    }
}
```

程序运行效果如图6-12所示。

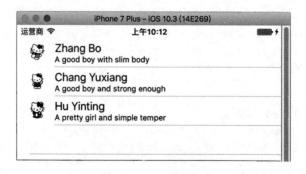

图6-12 一般表格效果

6.1.3 自定义表视图的使用

自定义表视图实际上就是指对单元格自定义,这样就可以出现和传统不一样的表格样式,如图 6-13 所示。

图 6-13 自定义单元格的表格

为了自定义单元格,需要新建一个类 MyCell,继承自 UITableViewCell,重载其初始化函数 override init(style:UITableViewCellStyle,reuseIdentifier:String?),在该初始化函数中,重新设计了不同位置的两个 UILabel 和一个 UIImageView;重载了 frame 变量,使得单元格宽度始终为屏幕宽度。

程序代码如下:

```
//  ViewController.swift
//  MyCellTableView
//  Created by Zhifeng Chen on 2017/6/4.
//  Copyright 2017年 Zhifeng Chen. All rights reserved.
import UIKit
class MyCell : UITableViewCell {
    //UI variable
    var nameLabel : UILabel!
    var characterLabel : UILabel!
    var headImageView : UIImageView!
```

```swift
        override init(style: UITableViewCellStyle, reuseIdentifier: String?) {
            super.init(style: style, reuseIdentifier: reuseIdentifier)
            //建立显示name的UILabel组件
            let nameRect = CGRect(x: 5, y: 5, width: self.frame.width - 5, height: self.frame.height/2)
            nameLabel = UILabel(frame: nameRect)
            nameLabel.font = UIFont.boldSystemFont(ofSize: 20)
            self.addSubview(nameLabel)
            //建立显示character的UILabel组件
            let characterRect = CGRect(x: 5, y: self.frame.height/2 + 20, width: self.frame.width - 5, height: self.frame.height/2)
            characterLabel = UILabel(frame: characterRect)
            self.addSubview(characterLabel)
            //建立显示头像照片headpic的UIImageView组件
            let headRect = CGRect(x: self.frame.width - self.frame.height, y: 5, width: self.frame.height, height: self.frame.height)
            headImageView = UIImageView(frame: headRect)
            self.addSubview(headImageView)
        }
        required init?(coder aDecoder: NSCoder) {
            fatalError("init(coder:) has not been implemented")
        }
        //让单元格宽度始终为屏幕宽
        override var frame: CGRect {
            get {
                return super.frame
            }
            set (newFrame) {
                var frame = newFrame
                frame.size.width = UIScreen.main.bounds.width
                super.frame = frame
            }
        }
    }
    class ViewController: UIViewController,UITableViewDelegate,UITableViewDataSource {
        //表格每个单元格显示的内容数组
        var dataArray : Array<Dictionary<String,String>> = [
            ["name":"Zhang Bo","character":"A good boy with slim body","headpic":"head01.png"],
            ["name":"Chang Yuxiang","character":"A good boy and strong enough","headpic":"head02.png"],
            ["name":"Hu Yinting","character":"A pretty girl and simple temper","headpic":"head03.png"]
        ]
        //建立一个myTablview的表格变量
        var myTableView : UITableView!
        override func viewDidLoad() {
            super.viewDidLoad()
            // Do any additional setup after loading the view, typically from a nib.
            //建立一个和屏幕大小一致的矩形框
            let rect = CGRect(x: 0, y: 20, width: self.view.frame.size.width, height: self.view.frame.size.height - 20)
```

```
        //根据矩形框大小生成一个表格
        myTableView = UITableView(frame: rect)
        //设置表格的代理
        myTableView.delegate = self
        //设置表格的数据代理
        myTableView.dataSource = self
        //显示表格
        self.view.addSubview(myTableView)
    }
    //Mark——tableView Datasource
    //确定表格每个分组有多少行?默认分组为1个
    func tableView(_ tableView: UITableView, numberOfRowsInSection section: Int) -> Int {
        return dataArray.count
    }
    //确定表格每个单元格的样式,和每个单元格显示的内容
    func tableView(_ tableView: UITableView, cellForRowAt indexPath: IndexPath) -> UITableViewCell {
        //根据MyCell来建立每个单元格"Cell"
        let cell = MyCell(style: UITableViewCellStyle.default, reuseIdentifier: "Cell")
        let dic = dataArray[indexPath.row]
        cell.nameLabel.text = dic["name"]
        cell.characterLabel.text = dic["character"]
        cell.headImageView.image = UIImage(named: dic["headpic"]!)
        return cell
    }
    //将每个单元格的高度设置为80
    func tableView(_ tableView: UITableView, heightForRowAt indexPath: IndexPath) ->CGFloat {
        return 80
    }
}
```

6.2 图灵机器人 API

北京××科技有限公司主要从事机器人人工智能及机器人操作系统的研发及商业化应用,在语义理解、机器视觉、多模态人机交互、深度学习、机器人等领域具备领先优势。

2014年11月发布图灵机器人(网址 http://www.tuling123.com),是中文语境下智能度较高的机器人大脑,已为超过23万家企业和开发者提供服务,广泛应用于机器人、智能家居、智能车载、智能客服、可穿戴设备等众多场景。

6.2.1 数据交换格式 JSON

JSON(JavaScript Object Notation)是一种轻量级的数据交换格式。它基于 ECMAScript(w3c 制定的 JS 规范)的一个子集,采用完全独立于编程语言的文本格式来存储和表示数据。JSON 拥有简洁和清晰的层次结构,易于人阅读和编写,同时也易于机器解析和生成,并有效地提升网络传输效率,使得 JSON 成为理想的数据交换语言。

JSON 数据分为3种形式:对象、数组、值。

对象是一个无序的"'名称/值'对"集合。一个对象以"{"(左括号)开始,"}"(右括号)结束。每个"名称"后跟一个":"(冒号);"'名称/值'对"之间使用","(逗号)分隔,如图 6-14 所示。

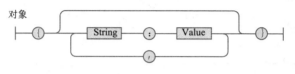

图 6-14 对象图示

数组是值(Value)的有序集合。一个数组以"["(左中括号)开始,"]"(右中括号)结束。值之间使用","(逗号)分隔,如图 6-15 所示。

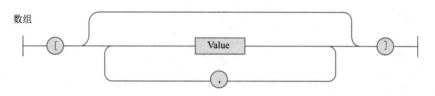

图 6-15 数组图示

值可以是双引号括起来的字符串(String)、数值(Number)、True、False、Null、对象(Object)或者数组(Array)。这些结构可以嵌套,如图 6-16 所示。

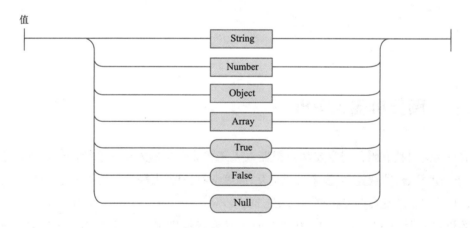

图 6-16 值图示

下面是一个简单的例子:

```
{
    "Name": "Chen Zhifeng",
    "Profession" : "Teacher",
    "Age": 46,
    "Email": "13402506301@ 163.com",
    "Friends": ["Zhang San","Li Si"]
}
```

上面的数据示例,表示了这样一个结构,首先数据被一对大括号包围,那么数据就是对象类型,然后它里面有 5 个属性:Name、Profession、Age、Email 和 Friends。其中 3 个属性 Name、Profession 和 Email 是字符串类型;Age 属性代表年龄,所以它的值是一个 Number 类型的 46。注意:字符串类型和数字类型的区别,字符串类型的值用一对双引号括了起来,而数值类型不需要双引号。

最后,Friends 属性的值是一个数组,用一对中括号包围起来,而数组中的元素,仍然是字符串类型。

下面再分析一个例子:某高校组建了一个 iOS 开发的兴趣小组,这个小组的名称为 iOS-Orange-Team,指导教师为 Chen Zhifeng,小组的学生有 3 个,分别为 Zhang bo、Chang Wenxiang 和 Hu yinting。

```
{
  "TeamName":"iOS-Orange-Team",
  "Teacher":"Chen Zhifeng",
  "Students":[
      {"Name": "Zhang Bo", "Hometown":"Xuzhou"},
      {"Name":" Chang Wenxiang ","Hometown":"Yancheng"},
      {"Name":" Hu Yinting","Hometown":"Suzhou"}
  ]
}
```

通过在浏览器中访问网址来获得英国伦敦的天气预报等信息:http://samples.openweathermap.org/data/2.5/weather？q=London,uk&appid=b1b15e88fa797225412429c1c50c122a1,得到类似以下 JSON 数据:

```
{
  "coord":{"lon":-0.13,"lat":51.51},
  "weather":[{"id":300,"main":"Drizzle","description":"light intensity drizzle","icon":"09d"}],
  "base":"stations",
  "main":{
    "temp":280.32,
    "pressure":1012,
    "humidity":81,
    "temp_min":279.15,
    "temp_max":281.15
  },
  "visibility":10000,
  "wind":{"speed":4.1,"deg":80},
  "clouds":{"all":90},
  "dt":1485789600,
  "sys":{
    "type":1,
    "id":5091,
    "message":0.0103,
    "country":"GB",
```

```
    "sunrise":1485762037,
    "sunset":1485794875
},
"id":2643743,
"name":"London",
"cod":200
}
```

6.2.2 数据测试与解析

1. 图灵机器人 JSON 数据测试

用户可以访问网址 http://www.tuling123.com，然后注册一个账号，可以免费访问 5000 次/天，其中最重要的是记住 APIKey，如图 6-17 所示。

图 6-17　图灵机器人注册成功后的登录界面

在通过 iOS 编程访问图灵机器人数据之前，可以通过浏览器测试一下网络是否通畅。在浏览器中输入以下网址，其中 Key =××××××××修改为刚才注册成功后获得的 APIKey，就可以获得如图 6-18 所示的 JSON 数据：

图 6-18　图灵机器人正常返回 JSON 数据

http://www.tuling123.com/openapi/api?key=××××××××&info=你是谁?&type=JSON

测试正常后,就可以通过编程来实现图灵机器人数据的获取。

2. Swift 中的 JSONSerialization

用户与 Web 应用通信,一般可以从服务端返回 JSON 格式的消息。因此,在 Swift 的 Foundation 框架中,提供了 JSONSerialization 类可以将 JSON 格式的数据转换为 Swift 的 Dictionary、Array、String、Number 和 Bool 等类型。当然,有时因无法确定 iOS 接收的 JSON 结构或值,也就无法正确地序列化对象模型。

【例6-1】首先建立一个字典 Dictionary,包括 3 个数据项,其中 Students 这个数据项是一个数组 Array(包括 3 个记录,每个记录也是一个字典 Dictionary),然后判断这个字典中的数据是否符合转换为 JSON 格式的合法性,如果数据格式正确,则调用 JSONSerialization.data 转换为 data,最后 JSONSerialization.jsonObject 将 data 中 JSON 格式重新解析为字典。

程序代码如下:

```swift
//  ViewController.swift
//  JsonDemo
//
//  Created by Zhifeng Chen on 2017/6/3.
//  Copyright 2017年 Zhifeng Chen. All rights reserved.

import UIKit
class ViewController: UIViewController {
    override func viewDidLoad() {
        super.viewDidLoad()
        // Do any additional setup after loading the view, typically from a nib.
        let sourceBody : [String : Any] = [
            "TeamName" : "iOS-Orange-Team",
            "Teacher":"Chen Zhifeng",
            "Students":[
                ["Name":"Zhang Bo" , "Hometown":"Xuzhou"],
                ["Name" : "Chang Wenxiang" , "Hometown":"Yancheng"],
                ["Name":"Hu Yinting" , "Hometown":"Suzhou"]
            ]
        ]
        //判断能否将 sourceBody 这个 Dictionary 中的数据转化为 JSON 格式
        if ! JSONSerialization.isValidJSONObject(sourceBody) {
            print("不能转化")
            return
        }
        //sourceBody 字符串被转换为 data 类型(JSON)
        guard let data = try? JSONSerialization.data(withJSONObject: sourceBody, options: .prettyPrinted) else {
            return
        }
        //从 data 类型转为 JSON 字符串
        let jsonstr = String(data: data, encoding: .utf8)
```

```swift
            print("Json Str" + jsonstr!)
            //.allowFragment 允许json字符串最外层不是array或dictionary 但必须是有效的json格式
            //.mutableContainers 返回可变容器array或dictionary
            if let json = try? JSONSerialization.jsonObject(with: data, options: .allowFragments) as! [String : Any] {
                
                if let TeamName = json["TeamName"] {
                    print(TeamName)
                }
                if let Teacher = json["Teacher"] {
                    print(Teacher)
                }
                if let Students = json["Students"] {
                    print(Students)
                }
            }
        }
    override func didReceiveMemoryWarning() {
        super.didReceiveMemoryWarning()
        // Dispose of any resources that can be recreated.
    }
}
```

程序运行结果如图6-19所示。

```
Json Str{
  "TeamName" : "iOS-Orange-Team",
  "Students" : [
    {
      "Name" : "Zhang Bo",
      "Hometown" : "Xuzhou"
    },
    {
      "Name" : "Chang Wenxiang",
      "Hometown" : "Yancheng"
    },
    {
      "Name" : "Hu Yinting",
      "Hometown" : "Suzhou"
    }
  ],
  "Teacher" : "Chen Zhifeng"
}
iOS-Orange-Team
Chen Zhifeng
(
        {
        Hometown = Xuzhou;
        Name = "Zhang Bo";
    },
        {
        Hometown = Yancheng;
        Name = "Chang Wenxiang";
    },
        {
        Hometown = Suzhou;
        Name = "Hu Yinting";
    }
)
```

图6-19 字典数据与JSON相互转换

6.3 网络访问 URLSession

NSURLSession 是苹果提供的原生网络访问类,提供了配置每个会话的缓存、协议、cookie 和证书政策,甚至跨应用程序共享它们的能力。这使得框架的网络基础架构和部分应用程序独立工作,而不会互相干扰。每一个 NSURLSession 对象都是根据一个 NSURLSessionConfiguration 初始化的,该 NSURLSessionConfiguration 指定策略,以及一系列为了提高移动设备性能而专门添加的新选项。

NSURLSession 的另一重要组成部分是会话任务,它负责处理数据的加载,以及客户端与服务器之间的文件和数据的上传下载服务。

需要注意的是,由于 NSURLSession 采用的是"异步阻塞"模型,所以在实现代理方法更新 UI 时需要将线程切回主线程。

6.3.1 NSURLSession 的用法

NSURLSession 的使用非常简单,先根据会话对象创建一个请求 Task,然后执行该 Task 即可。NSURLSessionTask 本身是一个抽象类,在使用的时候,通常是根据具体的需求使用它的几个子类,如图 6-20 所示。

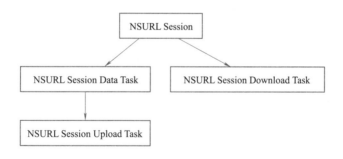

图 6-20 网络访问 NSURLSession 子类

NSURLSession 使用的第 1 步需要配置会话模式,其支持以下 3 种会话模式:

(1)默认会话模式(Default):默认模式,基于磁盘缓存的持久化策略,使用用户 keychain 中保存的证书进行认证授权。

(2)瞬时会话模式(Ephemeral):不存储任何数据在磁盘中,所有数据都保存在 RAM 中,当会话结束后,缓存数据将被清空。

(3)后台会话模式(Background):该模式类似于默认模式,只是将上传和下载移至后台处理,需要一个提供一个 String 用于标识后台会话。

第 2 步,在配置完会话模式后,就可以获取 NSURLSession 对象。获取对象的方法有以下几种:

(1) sharedSession 获取的会话使用的是默认配置(Default)，全局共享的 Cookies、Cache 和证书。

(2) 使用构造器构造一个指定配置的会话对象。

(3) 使用构造器构造一个指定配置对象，并指定代理及代理列队。

第 3 步，在获取完会话对象后，就需要设置会话任务。在这里是通过建立一个会话任务对象来实现布置任务的。在一个会话中，NSURLSession 支持 3 种会话任务：

(1) 数据任务(NSURLSessionDataTask)。

(2) 上传任务(NSURLSessionUploadTask)。

(3) 下载任务(NSURLSessionDownloadTask)。

第 4 步，最后获得任务对象后，就可以对它进行如下操作：

```
let dataTask = session.dataTask(with: request,
    completionHandler: {(data, response, error) ->Void in
        if error! = nil{
            print(error.debugDescription)
        }else{
            let str = String(data: data!, encoding: String.Encoding.utf8)
            print(str)
        }
}) as URLSessionTask
```

在实际使用时，常采用默认值来简化上述步骤。

1. 获取数据和下载文件

Data Task 加载数据：使用全局的 URLSession.shared 和 dataTask 方法创建。使用 NSData 对象来发送和接收数据。数据任务可以分片返回数据，也可以通过完成处理器一次性返回数据。由于数据任务不存储数据到文件，所以不支持后台会话。

程序代码如下：

```
func sessionGetData(){
    //创建 URL 对象
    let urlString = "http://www.tuling123.com/"
    let url = URL(string:urlString)
    //创建请求对象
    let request = URLRequest(url: url!)
    let session = URLSession.shared
    let dataTask = session.dataTask(with: request,
        completionHandler: {(data, response, error) ->Void in
            if error ! = nil{
                print(error.debugDescription)
            }else{
                let str = String(data: data!, encoding: String.Encoding.utf8)
                print(str!)
            }
    })as URLSessionTask
```

```
        //使用 resume 方法启动任务
        dataTask.resume()
}
```

注意:苹果要求 APP 内访问的网络必须使用 HTTPS 协议,为了能在 iOS 中访问 http 网络数据,必须在 Xcode 的工程文件中,找到 Info.plist 文件,在里面添加相关键值 App Transport Security Setting→Allow Arbitrary Loads→YES,如图 6-21 所示。

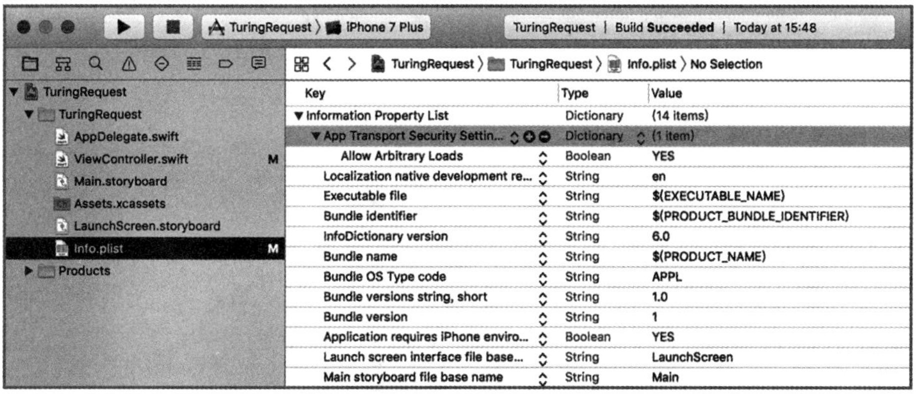

图 6-21　开放 http 网络访问设置

如果 Info.plist 文件设置不正确,会出现提示提示信息:App Transport Security has blocked a cleartext HTTP (http://) resource load since it is insecure. Temporary exceptions can be configured via your app's Info.plist file。

Download Task 下载文件:以文件的形式接收数据,当程序不运行时支持后台下载。使用全局的 URLSession.shared 和 dataTask 方法创建。通过下载指定的图片文件到应用程序的 Documents 目录中,采用了时间戳,保证文件不会重名。

程序代码如下:

```
func sessionDownloadImage(){
    //下载地址
    let url = URL(string: "http://hangge.com/blog/images/logo.png")
    //请求
    let request = URLRequest(url: url!)
    let session = URLSession.shared
    //下载任务
    let downloadTask = session.downloadTask(with: request,completionHandler:
    { (location:URL?, response:URLResponse?, error:Error?) ->Void in
        
        //输出下载文件原来的存放目录
        print("location:\\(String(describing: location))")
        //location 位置转换
        let locationPath = location?.path
        //获取当前时间
        let now = NSDate()
```

```
        //当前时间的时间戳
        let timeInterval:TimeInterval = now.timeIntervalSince1970
        let timeStamp = String(timeInterval)
        //复制到用户目录
        let documents:String = NSHomeDirectory() + "/Documents/\\(timeStamp).png"
        //创建文件管理器
        let fileManager = FileManager.default
        try! fileManager.moveItem(atPath: locationPath!, toPath: documents)
        print("new location:\\(documents)")
    })

    //使用 resume()方法启动任务
    downloadTask.resume()
}
```

2. 服务器设置和文件上传

为了服务器能接收用户发送的文件,需要在自己的 Mac 计算机进行相关的设置。macOS Sierra 不但内置 Apache 服务器,还包括 PHP、Python、Ruby、Perl 等常用的脚本语言,这些都不需要用户自己编译安装,只需开启 Apache 和支持 PHP 即可使用。

图 6-22 所示为 Mac 中 Apache 服务器根目录/Library/Web Server/Documents。

图 6-22　Mac 中 Apache 服务器根目录/Library/WebServer/Documents

可以通过如下命令进行开启、关闭及重启:

```
$ sudo apachectl start | stop | restart
```

开启后,打开浏览器,访问 http://localhost/如果出现 It works!,则 Apache 可以正常使用。

macOS Sierra 已内置了 PHP 5.6，因此只需要在 Apache 的配置中加载 PHP 模块即可。

打开 Apache 配置文件/etc/apache2/httpd.conf，找到如下代码，去掉前面的注释（#）：

```
#LoadModule php5_module libexec/apache2/libphp5.so
```

默认没有生成 php.ini 配置文件，运行如下命令生成，也可以直接复制改名字：

```
$ sudo cp /etc/php.ini.default /etc/php.ini
```

重启 Apache 后，在 /Users/sean/webroot 目录下新建 phpinfo.php，内容如下：

```
<?php
phpinfo();
?>
```

打开浏览器，访问 http://localhost/phpinfo.php，如果出现 PHP 的相关信息，则配置成功。

在服务器根文件夹中建立一个 uploadFiles 子文件夹，然后配置能接收文件上传的服务器端程序 uploadSwift.php。

程序代码如下：

```php
<?php
/**  php接收流文件
 * @param String  $file 接收后保存的文件名
 * @return boolean
 */
function receiveStreamFile($receiveFile){
    $streamData = isset($GLOBALS['HTTP_RAW_POST_DATA'])? $GLOBALS['HTTP_RAW_POST_DATA'] : '';
    if(empty($streamData)){
        $streamData = file_get_contents('php://input');
    }
    if($streamData!=''){
        $ret = file_put_contents($receiveFile, $streamData, true);
    }else{
        $ret = false;
    }
    return $ret;
}
//定义服务器存储路径和文件名
$receiveFile = $_SERVER["DOCUMENT_ROOT"]."/uploadFiles/swift.png";
echo $receiveFile;
$ret = receiveStreamFile($receiveFile);
echo json_encode(array('success'=>(bool)$ret));
?>
```

Upload Task 上传文件：通常以文件的形式发送数据，支持后台上传。

程序代码如下:

```swift
func sessionUploadPhp(){
    //上传地址
    let url = URL(string: "http://localhost/uploadSwift.php")
    //1.创建会话对象
    let session = URLSession.shared
    //请求
    var request = URLRequest(url: url!,
    cachePolicy:.reloadIgnoringCacheData)
    request.httpMethod = "POST"
    //上传数据流
    let fileImage = Bundle.main.path(forResource: "bee1", ofType: "png")
    let imgData = try! Data(contentsOf: URL(fileURLWithPath: fileImage!))
    let uploadTask = session.uploadTask(with: request as URLRequest, from: imgData) {
        (data:Data?, response:URLResponse?, error:Error?) ->Void in
        //上传完毕后
        if error ! =nil{
            print(error!)
        }else{
            let str = String(data: data!, encoding: String.Encoding.utf8)
            print("上传完毕:\\(String(describing: str))")
        }
    }
    
    //使用 resume 方法启动任务
    uploadTask.resume()
}
```

6.3.2 图灵机器人网络数据访问

HTTP 定义了与服务器交互的不同方法,最基本的方法有 4 种,分别是 GET、POST、PUT、DELETE。URL 全称是资源描述符,可以这样认为:一个 URL 地址,用于描述一个网络上的资源,而 HTTP 中的 GET、POST、PUT、DELETE 就对应着对这个资源的查、改、增、删 4 个操作。因此,GET 一般用于获取/查询资源信息,而 POST 一般用于更新资源信息。

因为 GET 是通过 URL 提交数据,那么 GET 可提交的数据量就跟 URL 的长度有直接关系。而实际上,URL 不存在参数上限的问题,HTTP 协议规范没有对 URL 长度进行限制。这个限制是特定的浏览器及服务器对它的限制。IE 对 URL 长度的限制是 2 083 B。对于其他浏览器,如 Netscape、FireFox 等,理论上没有长度限制,其限制取决于操作系统的支持。

POST 把提交的数据放置在是 HTTP 包的包体中。理论上讲,POST 数据是没有大小限制的,HTTP 协议规范也没有进行大小限制,对 POST 数据起限制作用的是服务器处理程序的能力。

下面分别采用 GET 方法和 POST 方法访问图灵机器人,获取机器人回复的信息。

1. 采用 get 方法获取信息

GET 请求的数据会附在 URL 之后(就是把数据放置在 HTTP 协议头中),以"?"分割 URL 和传输数据,参数之间以 & 相连。

例如:

http://www.tuling123.com/openapi/api? key=3942&info=%E4%BD%A0%E5%A5%BD

如果数据是英文字母/数字,原样发送,如果是空格,转换为%20,如果是中文/其他字符,则直接把字符串用 BASE64 加密,得出如:%E4%BD%A0%E5%A5%BD,其中%××中的××为该符号以十六进制表示的 ASCII。

用户采用前面通过 NSURLSession 获取服务器数据的方法,稍做改动,即可从服务器获得回答。

程序代码如下:

```
//获取图灵机器人信息
func turingGet(question : String) {
    //创建 URL 对象
    //let str: String = "http://www.tuling123.com/openapi/api?"
    //let queryItem1 = NSURLQueryItem(name: "info", value: question) as URLQueryItem
    //let queryItem2 = NSURLQueryItem( name: " key ", value: "3942ce7f92074d97b02efed31e37e487") as URLQueryItem
    //let urlCom : NSURLComponents! = NSURLComponents(string: str)
    //urlCom.queryItems = [queryItem1 , queryItem2]
    //let url = urlCom.url!

    let str = "http://www.tuling123.com/openapi/api? key=3942ce7f92074d97b02efed31e37e487&info=\\(question)"
    let cs = NSCharacterSet(charactersIn:"").inverted
    let geturl = str.addingPercentEncoding(withAllowedCharacters: cs)
    let url = URL(string: geturl!)
    //创建请求对象
    let request = URLRequest(url: url!)
    //创建会话 Session
    let session = URLSession.shared
    //创建任务 dataTask
```

```
        let dataTask = session.dataTask(with: request,
            completionHandler: {(data, response, error) ->Void in
                if error ! =nil{
                    print(error.debugDescription)
                }else{
                    let str = String(data: data!, encoding: String.Encoding.utf8)
                    print(str!)
                }
        }) as URLSessionTask

        //使用 resume 方法启动任务
        dataTask.resume()
}
```

采用 get 方法获取信息的运行效果如图 6-23 所示。

图 6-23　采用 get 方法获取信息的运行效果

2. 采用 post 方法获取信息

POST 把提交的数据则放置在是 HTTP 包的包体中,安全性相对要高一些。用户采用前面通过 NSURLSession 获取服务器数据的 get 方法,稍做改动,即可从服务器获得回答。

程序代码如下：

```
//获取图灵机器人信息
func turingPost(question : String) {
    let str ="http://www.tuling123.com/openapi/api"
    let url: NSURL = NSURL(string: str)!
    //创建请求对象
    let request: NSMutableURLRequest = NSMutableURLRequest(url: url as URL)
    //修改请求方法为 POST
    request.httpMethod = "POST"
    //创建会话 Session
    let session = URLSession.shared
```

```
        //设置请求体
        request.httpBody = "key=3942ce7f92074d97b02efed31e37e487&info=\\(question)".data(using: String.Encoding.utf8)
        let dataTask: URLSessionDataTask = session.dataTask(with: request as URLRequest) { (data, response, error) in
            //解析数据
            let str = String(data: data!, encoding: String.Encoding.utf8)
            print(str!)
        }
        //使用 resume 方法启动任务
        dataTask.resume()
    }
```

6.4 基于表格的聊天界面

6.4.1 简单表格聊天界面

下面把前面学习的知识组合起来使用,实现一个最简单的表视图来显示聊天界面,采用 POST 方法来向机器人提问,并获得回答,将机器人的回答进行 JSON 数据分析,获得其内容,然后通过一个主线程将数据传送给表格显示出来。

程序代码如下:

```
//  ViewController.swift
//  TableTuringChat
//  Created by Zhifeng Chen on 2017/6/4.
//  Copyright 2017年 Zhifeng Chen. All rights reserved.
import UIKit
class ViewController: UIViewController ,UITableViewDataSource,UITableViewDelegate {
    //variable
    //表格每个单元格显示的内容数组
    var dataArray : Array<String> = []
    //建立一个 myTablview 的表格变量
    var myTableView : UITableView!
    //向机器人提问内容
    var question : Array<String> = ["苏州天气如何?","你是谁啊?","你知道星港学校吗?"]
    override func viewDidLoad() {
        super.viewDidLoad()
        // Do any additional setup after loading the view, typically from a nib.
        //建立一个和屏幕大小一致的矩形框
        let rect = CGRect(x: 0, y: 20, width: self.view.frame.size.width, height: self.view.frame.size.height-20)
        //根据矩形框大小生成一个表格
        myTableView = UITableView(frame: rect)
        //设置表格的代理
        myTableView.delegate = self
```

```swift
        //设置表格的数据代理
        myTableView.dataSource = self
        //显示表格
        self.view.addSubview(myTableView)

        for each in question {
        turingPost(question: each)
        }
    }
    //获取图灵机器人信息
    func turingPost(question : String) {
        let str = "http://www.tuling123.com/openapi/api"
        let url: NSURL = NSURL(string: str)!
        //创建请求对象
        let request: NSMutableURLRequest = NSMutableURLRequest(url: url as URL)
        //修改请求方法为 POST
        request.httpMethod = "POST"
        //创建会话 Session
        let session = URLSession.shared
        //设置请求体
        request.httpBody = "key=3942ce7f92074d97b02efed31e37e487&info=\\(question)".data(using: String.Encoding.utf8)
        let dataTask: URLSessionDataTask = session.dataTask(with: request as URLRequest) { (data, response, error) in
            //解析数据
            if let json = try? JSONSerialization.jsonObject(with: data!, options: .allowFragments) as! [String : Any] {
                let content : String = json["text"] as! String
                DispatchQueue.main.async(execute: { () -> Void in
                    self.dataArray.append("我问:\\(question)")
                    self.dataArray.append("机器人答:\\(content)")
                    self.myTableView.reloadData()
                })
            }
        }
        //使用 resume 方法启动任务
        dataTask.resume()
    }
    //Mark——tableView Datasource
    //确定表格每个分组有多少行? 默认分组为1个
    func tableView(_ tableView: UITableView, numberOfRowsInSection section: Int) -> Int {
        return dataArray.count
    }
    //确定表格每个单元格的样式,和每个单元格显示的内容
    func tableView(_ tableView: UITableView, cellForRowAt indexPath: IndexPath) -> UITableViewCell {
        let cell = UITableViewCell(style: UITableViewCellStyle.default, reuseIdentifier: "Cell")
        let title = dataArray[indexPath.row]
        cell.textLabel?.text = title
        return cell
    }
}
```

程序运行效果如图 6-24 所示。

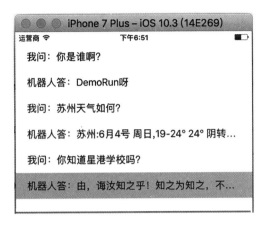

图 6-24　程序运行效果

6.4.2　一般表格聊天界面

同简单表格聊天界面类似,采用 POST 方法来向机器人提问,并获得回答,将机器人的回答进行 JSON 数据分析,获得其内容,整合到一个 Dictionary 中,然后通过一个主线程将数据传送给表格显示出来。

程序代码如下:

```
//  ViewController.swift
//  TableTuringChat
//  Created by Zhifeng Chen on 2017/6/4.
//  Copyright 2017 年 Zhifeng Chen. All rights reserved.
import UIKit
class ViewController: UIViewController ,UITableViewDataSource,UITableViewDelegate {
    //variable
    //表格每个单元格显示的内容数组
    var dataArray : Array<Dictionary<String,String>> = []
    //建立一个 myTablview 的表格变量
    var myTableView : UITableView!
    //向机器人提问内容
    var question : Array<String> = ["苏州天气如何?","你是谁啊?","你知道星港学校吗?","China 是什么意思?"]
    override func viewDidLoad() {
        super.viewDidLoad()
        // Do any additional setup after loading the view, typically from a nib.
        //建立一个和屏幕大小一致的矩形框
        let rect = CGRect(x: 0, y: 20, width: self.view.frame.size.width, height: self.view.frame.size.height -20)
        //根据矩形框大小生成一个表格
        myTableView = UITableView(frame: rect)
        //设置表格的代理
        myTableView.delegate = self
        //设置表格的数据代理
```

```swift
        myTableView.dataSource = self
        //显示表格
        self.view.addSubview(myTableView)
        for each in question {
            turingPost(question: each)
        }
    }
    //获取图灵机器人信息
    func turingPost(question : String) {
        let str = "http://www.tuling123.com/openapi/api"
        let url: NSURL = NSURL(string: str)!
        //创建请求对象
        let request: NSMutableURLRequest = NSMutableURLRequest(url: url as URL)
        //修改请求方法为 POST
        request.httpMethod = "POST"
        //创建会话 Session
        let session = URLSession.shared
        //设置请求体
        request.httpBody = "key=3942ce7f92074d97b02efed31e37e487&info=\\(question)".data(using: String.Encoding.utf8)
        let dataTask: URLSessionDataTask = session.dataTask(with: request as URLRequest) { (data, response, error) in
            //解析数据
            if let json = try? JSONSerialization.jsonObject(with: data!, options: .allowFragments) as! [String : Any] {
                let content : String = json["text"] as! String
                let code = json["code"] as! Int
                DispatchQueue.main.async(execute: { () ->Void in
                    let myDic : Dictionary < String, String >=[
                    "who":"me","content":"\\(question)","code":"3266","headpic":"head01.png"
                    ]
                    self.dataArray.append(myDic)
                    let robotDic : Dictionary < String, String > = [
                    "who":"robot","content":"\\(content)","code":"\\(code)","headpic":"head03.png"
                    ]
                    self.dataArray.append(robotDic)
                    self.myTableView.reloadData()
                })
            }
        }
        //使用 resume 方法启动任务
        dataTask.resume()
    }
    
    //Mark——tableView Datasource
    //确定表格每个分组有多少行？默认分组为 1 个
    func tableView(_ tableView: UITableView, numberOfRowsInSection section: Int) -> Int {
        return dataArray.count
    }
```

```
//确定表格每个单元格的样式,和每个单元格显示的内容
func tableView(_ tableView: UITableView, cellForRowAt indexPath: IndexPath) -> UITableViewCell {
    let cell = UITableViewCell(style: UITableViewCellStyle.subtitle, reuseIdentifier: "Cell")
    let dic = dataArray[indexPath.row]
    let who = dic["who"]
    if who == "me" {
        cell.textLabel?.text = dic["content"]
    }
    else if who == "robot" {
        cell.detailTextLabel?.text = dic["content"]
    }
    cell.imageView?.image = UIImage(named: dic["headpic"]!)
    return cell
}
```

程序运行结果如图6-25所示。

图6-25 一般表格的机器人聊天界面

6.4.3 自定义表格聊天界面

通过自定义单元格样式,形成了全新的聊天界面,如图6-26所示。

本程序依然采用POST方法来向机器人提问,并获得回答,将机器人的回答进行JSON数据分析,获得其内容,整合到一个Dictionary中,然后通过一个主线程将数据传送给表格显示出来。

程序代码如下：

```swift
// ViewController.swift
// TableTuringChat
// Created by Zhifeng Chen on 2017/6/4.
// Copyright 2017 年 Zhifeng Chen. All rights reserved.
import UIKit
class MyCell : UITableViewCell {
    //UI variable
    var myLabel : UILabel!
    var robotLabel : UILabel!
    var myImageView : UIImageView!
    var robotImageView : UIImageView!
    override init(style: UITableViewCellStyle, reuseIdentifier: String?) {
        super.init(style: style, reuseIdentifier: reuseIdentifier)
        //建立显示 myLabel 的 UILabel 组件
        let myRect = CGRect(x: self.frame.width - self.frame.width/3*2 , y: 5, width: self.frame.width/2, height: self.frame.height-10)
        myLabel = UILabel(frame: myRect)
        myLabel.font = UIFont.boldSystemFont(ofSize: 20)
        myLabel.numberOfLines = 0
        myLabel.textAlignment = NSTextAlignment.right
        myLabel.adjustsFontSizeToFitWidth = true
        self.addSubview(myLabel)
        //建立显示 robotLabel 的 UILabel 组件
        let robotRect = CGRect(x: 60, y: 5 , width: self.frame.width/3*2, height: self.frame.height-10)
        robotLabel = UILabel(frame: robotRect)
        robotLabel.numberOfLines = 0
        robotLabel.textAlignment = NSTextAlignment.left
        robotLabel.adjustsFontSizeToFitWidth = true
        self.addSubview(robotLabel)
        //建立显示头像照片 myImageView 的 UIImageView 组件
        let myImgRect = CGRect(x: self.frame.width - 50, y: 5,width: 50, height: 50)
        myImageView = UIImageView(frame: myImgRect)
        self.addSubview(myImageView)
        //建立显示头像照片 robotImageView 的 UIImageView 组件
        let robotImgRect = CGRect(x: 5, y: 5, width: 50, height: 50)
        robotImageView = UIImageView(frame: robotImgRect)
        self.addSubview(robotImageView)
    }
    required init?(coder aDecoder: NSCoder) {
        fatalError("init(coder:) has not been implemented")
    }
    //让单元格宽度始终为屏幕宽
    override var frame: CGRect {
        get {
```

图 6-26　自定义单元格的聊天界面

```swift
            return super.frame
        }
        set (newFrame) {
            var frame = newFrame
            frame.size.width = UIScreen.main.bounds.width
            super.frame = frame
        }
    }
}
class ViewController: UIViewController ,UITableViewDataSource,UITableViewDelegate {
    //variable
    //表格每个单元格显示的内容数组
    var dataArray : Array<Dictionary<String,String>> = []
    //建立一个myTablview的表格变量
    var myTableView : UITableView!
    //向机器人提问内容
    var question : Array<String> = ["苏州天气如何?","你是谁啊?","你知道星港学校吗?","China 是什么意思?"]
    override func viewDidLoad() {
        super.viewDidLoad()
        // Do any additional setup after loading the view, typically from a nib.
        //建立一个和屏幕大小一致的矩形框
        let rect = CGRect(x: 0, y: 20, width: self.view.frame.size.width, height: self.view.frame.size.height - 20)
        //根据矩形框大小生成一个表格
        myTableView = UITableView(frame: rect)
        //设置表格的代理
        myTableView.delegate = self
        //设置表格的数据代理
        myTableView.dataSource = self
        //显示表格
        self.view.addSubview(myTableView)
        for each in question {
            turingPost(question: each)
        }
    }
    //获取图灵机器人信息
    func turingPost(question : String) {
        let str = "http://www.tuling123.com/openapi/api"
        let url: NSURL = NSURL(string: str)!
        //创建请求对象
        let request: NSMutableURLRequest = NSMutableURLRequest(url: url as URL)
        //修改请求方法为 POST
        request.httpMethod = "POST"
        //创建会话 Session
        let session = URLSession.shared
        //设置请求体
        request.httpBody = "key=3942ce7f92074d97b02efed31e37e487&info=\(question)".data(using: String.Encoding.utf8)
```

```swift
            let dataTask: URLSessionDataTask = session.dataTask(with: request as URLRequest) { (data, response, error) in
                //解析数据
                if let json = try? JSONSerialization.jsonObject(with: data!, options: .allowFragments) as! [String : Any] {
                    let content : String = json["text"] as! String
                    let code = json["code"] as! Int
                    DispatchQueue.main.async(execute: { () -> Void in
                        let myDic : Dictionary<String,String> = [
                            "who":"me","content":"\\(question)","code":"3266","headpic":"head01.png"
                        ]
                        self.dataArray.append(myDic)
                        let robotDic : Dictionary<String,String> = [
                            "who":"robot","content":"\\(content)","code":"\\(code)","headpic":"head03.png"
                        ]
                        self.dataArray.append(robotDic)
                        self.myTableView.reloadData()
                    })
                }
            }
            //使用resume方法启动任务
            dataTask.resume()
        }
        //Mark——tableView Datasource
        //确定表格每个分组有多少行？默认分组为1个
        func tableView(_ tableView: UITableView, numberOfRowsInSection section: Int) -> Int {
            return dataArray.count
        }
        //确定表格每个单元格的样式,和每个单元格显示的内容
        func tableView(_ tableView: UITableView, cellForRowAt indexPath: IndexPath) -> UITableViewCell {
            let cell = MyCell(style: UITableViewCellStyle.default, reuseIdentifier: "Cell")
            let dic = dataArray[indexPath.row]
            let who = dic["who"]
            if who == "me" {
                cell.myLabel?.text = dic["content"]
                cell.myImageView?.image = UIImage(named: dic["headpic"]!)
            }
            else if who == "robot" {
                cell.robotLabel?.text = dic["content"]
                cell.robotImageView?.image = UIImage(named: dic["headpic"]!)
            }
            return cell
        }
        //将每个单元格的高度设置为80
        func tableView(_ tableView: UITableView, heightForRowAt indexPath: IndexPath) -> CGFloat {
            return 80
        }
    }
```

附录 A

用户界面要素

1. 应用程序图标尺寸 App Icon Sizes

每个应用程序都需要一个美丽的和令人难忘的图标,在应用程序商店或者主屏幕上吸引人们的关注。

应用程序安装后,系统会根据硬件的分辨率选择相应的图标在主屏幕上或整个系统中显示,因此,每一个应用程序必须提供小型和大型应用程序图标。为不同的设备提供不同大小的小图标,可以确保应用程序图标看起来完美,从而支持所有的设备。

应用程序图标的标准如表 A-1 所示。

表 A-1 应用程序图标的标准

Device or context	Icon size
iPhone	180px × 180px（60pt × 60pt @ 3x）
	120px × 120px（60pt × 60pt @ 2x）
iPad Pro	167px × 167px（83.5pt × 83.5pt @ 2x）
iPad, iPad mini	152px × 152px（76pt × 76pt @ 2x）
App Store	1 024px × 1 024px

虽然大图标的使用效果不同于小图标,不过它仍然是应用程序图标。一般在外观上要与较小的图片相匹配,能展现更微妙、更丰富和更详细的视觉效果。

2. 突出显示(Spotlight)、设置(Settings)和消息通知(Notification)图标

每个应用程序还应该在搜索匹配时候提供一个小图标用于显示。另外,在应用程序设

置的时候应该内建提供一个小图标,一些支持消息通知的应用程序在获得通知的时候也需要提供一个小图标。所以,这些小图标都应该能清晰地标示应用程序,理想情况下,要和应用程序图标一致。当然,如果没有提供这些图标,iOS 会自动缩放主应用程序图标来满足这些显示的需要。

Spotlight、Settings、Notification 的设置规范如表 A-2 所示。

表 A-2　Spotlight、Settings、Notification 设置规范

Device	Spotlight icon size
iPhone	120px×120px（40pt×40pt @3x）
	80px×80px（40pt×40pt @2x）
iPad Pro、iPad、iPad mini	80px×80px（40pt×40pt @2x）
Device	Settings icon size
iPhone	87px×87px（29pt×29pt @3x）
	58px×58px（29pt×29pt @2x）
iPad Pro、iPad、iPad mini	58px×58px（29pt×29pt @2x）
Device	Notification icon size
iPhone	60px×60px（20pt×20pt @3x）
	40px×40px（20pt× 20pt @2x）
iPad Pro、iPad、iPad mini	40px×40px（20pt×20pt @2x）

3. 自定义图标(Custom Icons)

如果系统提供的图标还不能满足应用程序所包括的任务或者模式,或者系统图标和应用程序风格不匹配,就应该建立自己的图标。一个自定义的图标,有时候称为模版,颜色等信息会被丢弃,只使用其轮廓,常常被用于导航条、tab、工具栏或者屏幕快速操作等。图 A-1 所示为部分自定义图标。

图 A-1　部分自定义图标

简洁而富有特点的设计:太多的细节有时候会让图标很不容易理解。能被大众很好地理解是最重要的。

设计透明、抗锯齿、无阴影的单一颜色图标:iOS 会忽略颜色信息,因此没有必要使用多种颜色,透明有利于突出图标形状。

图标的一致性:无论是只使用自定义图标,或者混合使用系统图标,所有的图标都应该使用相同规格的尺寸、一致的细节表现,包括笔画粗细等。如果想让图标和 iOS 的相近,建议使用细笔画,一般在 1pt(在@2x 精度采用 2px)。

提供两种版本的 tab 图标：提供选中和没有选中两种状态下的不同图标，如图 A-2 所示。一般来说，填充的图标用于选中。

在 tab 上不要使用文字：如果需要显示文字，建议在下方与图标一致的位置。自定义图标尺寸如表 A-3 所示。

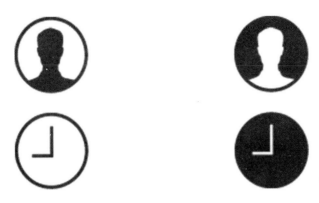

图 A-2　两种状态下的不同图标

表 A-3　自定义图标尺寸

图标尺寸值	Navigation bar and toolbar icon size	Tab bar icon size
推荐值	75px×75px（25pt×25pt @3x）	
	50px×50px（25pt×25pt @2x）	
最大值	83px×83px（27.67pt×27.67pt @3x）	144px×96px（48pt×32pt @3x）
	56px×56px（28pt×28pt @2x）	96px×64px（48pt×32pt @2x）

4. 静态启动画面（Static Launch Screen Images）

设置参数如表 A-4 所示。

表 A-4　静态启动画设置参数

Device	Portrait size	Landscape size
iPhone 7 Plus，iPhone 6s Plus	1080px×1920px	1920px×1080px
iPhone 7，iPhone 6s	750px×1334px	1334px×750px
iPhone SE	640px×1136px	1136px×640px
12.9-inch iPad Pro	2048px×2732px	2732px×2048px
9.7-inch iPad Pro，iPad Air 2，iPad mini 4，iPad mini 2	1536px×2048px	2048px×1536px

5. 系统图标（System Icons）（导航条和工具栏图标）

其名称及功能如表 A-5 所示。

表 A-5 系统图标(导航及和工具栏图标)名称及功能

Icon	Button name	Function
	Action	Shows a modal view containing share extensions, action extensions, and tasks, such as Copy, Favorite, or Find, that are useful in the current context.
	Add	Creates a new item.
	Bookmarks	Shows app-specific bookmarks.
	Camera	Takes a photo or video, or shows the Photo Library.
Cancel	Cancel	Closes the current view or ends edit mode without saving changes.
	Compose	Opens a new view in edit mode.
Done	Done	Saves the state and closes the current view, or exits edit mode.
Edit	Edit	Enters edit mode in the current context.
	Fast Forward	Fast-forwards through media playback or slides.
	Organize	Moves an item to a new destination, such as a folder.
	Pause	Pauses media playback or slides. Always store the current location when pausing, so playback can resume later.
	Play	Begins or resumes media playback or slides.
Redo	Redo	Redoes the last action that was undone.
	Refresh	Refreshes content. Use this icon sparingly, as your app should refresh content automatically whenever possible.
	Reply	Sends or routes an item to another person or location.

Icon	Button name	Function
⏪	Rewind	Moves backwards through media playback or slides.
Save	Save	Saves the current state.
🔍	Search	Displays a search field.
✕	Stop	Stops media playback or slides.
🗑	Trash	Deletes the current or selected item.
Undo	Undo	Undoes the last action.

6. 系统图标（System Icons）（Tab Bar 图标）

其名称、功能如表 A-6 所示。

表 A-6　系统图标（Tab Bar 图标）名称、功能

Icon	Button name	Function
📖	Bookmarks	Shows app – specific bookmarks.
👤	Contacts	Shows the person's contacts.
⬇	Downloads	Shows active or recent downloads.
☆	Favorites	Shows the person's favorite items.
★	Featured	Shows content featured by the app.
🕒	History	Shows recent actions or activity.
⋯	More	Shows additional tab bar items.
🕒	Most Recent	Shows the most recent items.

续表

Icon	Button name	Function
★▤	Most Viewed	Shows the most popular items.
🕒	Recents	Shows content or items recently accessed within a specific period of time.
🔍	Search	Enters a search mode.
★	Top Rated	Shows the highest-rated items.

7. 系统图标（System Icons）（Quick Action 图标）

其名称、功能如表 A-7 所示。

表 A-7　系统（Quick Action 图标）名称及功能

Icon	Button name	Function
＋	Add	Creates a new item.
⏰	Alarm	Sets or displays an alarm.
🔊	Audio	Denotes or adjusts audio.
📷	Capture Photo	Captures a photo.
📖	Bookmark	Creates a bookmark or shows bookmarks.
🎥	Capture Video	Captures a video.
☁	Cloud	Denotes, displays, or initiates a cloud-based service.
✎	Compose	Composes new editable content.
✓	Confirmation	Denotes that an action is complete.

Icon	Button name	Function
	Contact	Chooses or displays a contact.
	Date	Displays a calendar or event, or performs a related action.
	Favorite	Denotes or marks a favorite item.
	Home	Indicates or displays a home screen. Indicates, displays, or routes to a physical home.
	Invitation	Denotes or displays an invitation.
	Location	Denotes the concept of location or accesses the current geographic location.
	Love	Denotes or marks an item as loved.
	Mail	Creates a Mail message.
	Mark Location	Denotes, displays, or saves a geographic location.
	Message	Creates a new message or denotes the use of messaging.
	Pause	Pauses media playback. Always store the current location when pausing, so playback can resume later.
	Play	Begins or resumes media playback.
	Prohibit	Denotes that something is disallowed.
	Search	Enters a search mode.
	Share	Shares content with others or to social media.

续表

Icon	Button name	Function
⤨	Shuffle	Indicates or initiates shuffle mode.
○	Task	Denotes an uncompleted task or marks a task as complete.
◉	Task Completed	Denotes a completed task or marks a task as not complete.
🕒	Time	Denotes or displays a clock or timer.
⤓	Update	Updates content.

附录 B

iOS 俱乐部护照

OUR VISION

创造，人生第一个 iOS App
Create our First iOS App

照片

姓/Surname

名/Given names

护照号/Passport NO.:

学校/University　　学号/Student ID

指导老师/Teacher　　签发日期/Date of issue

iOS Club 活动签证
Activity Visa

每参加一次 iOS Club 社团活动，即可获得一枚活动签证印章，作为在 iOS Club 期间参加交流、分享和学习的证明。

活动签证　A Visas

盖章日期：　　　　　　　盖章日期：

盖章日期：　　　　　　　盖章日期：

iOS Club 活动签证
Activity Visa

每参加一次 iOS Club 社团活动，即可获得一枚活动签证印章，作为在 iOS Club 期间参加交流、分享和学习的证明。

活动签证　A Visas

盖章日期：　　　　　　　盖章日期：

盖章日期：　　　　　　　盖章日期：

iOS App 开发签证
Development Visa

在 iOS Club 期间，开发并提交一款 iOS App，即可获得一枚开发签证印章，作为完成 iOS Club 终极目标的证明。

开发签证　D Visas

盖章日期：　　　　　　盖章日期：

盖章日期：　　　　　　盖章日期：

---

iOS App 开发签证
Development Visa

在 iOS Club 期间，开发并提交一款 iOS App，即可获得一枚开发签证印章，作为完成 iOS Club 终极目标的证明。

开发签证　D Visas

盖章日期：　　　　　　盖章日期：

盖章日期：　　　　　　盖章日期：

iOS App 开发签证
Development Visa

在 iOS Club 期间，开发并提交一款 iOS App，即可获得一枚开发签证印章，作为完成 iOS Club 终级目标的证明。

开发签证　　D Visas

盖章日期：　　　　　盖章日期：

盖章日期：　　　　　盖章日期：

开发签证　　A Visas

盖章日期：　　　　　盖章日期：

盖章日期：　　　　　盖章日期：

使用守则

- iOS Club Passport 是 iOS Club 会员的身份凭证，用于记录会员在 iOS Club 期间的活动经历和开发成就，并凭此享有 iOS Club 的相关福利。
- 使用前请在首页粘贴个人照片，并填写个人真实信息，以便社团指导老师核实会员身份。
- 每次参加 iOS Club 社团活动时请务必携带本 Passport，并于活动结束后交由社团指导老师统一盖章和签署日期。
- 本 Passport 仅限一人使用，不得转借他人。如有遗失，原 Passport 的活动记录作废，请妥善保存。
- iOS Club Passport 仅供在校学生使用，毕业离校之后即自动失效。